DATE DUE			
OCT 18 1993			

HOW OLD IS THE EARTH?

More than twenty years ago PATRICK M. HURLEY read Arthur Holmes' little book *The Age of the Earth* and since has been unable to tear himself away from the subject. Geological age measurement and the application of nuclear physics to geology have been his special scientific interests, and *How Old Is the Earth?*, his first book, is a distillation of what he has learned about these fascinating subjects in his unusually varied career.

Born in Hong Kong, in 1912, he moved with his family to Vancouver Island, British Columbia, at the age of nine. He was graduated from the University of British Columbia with a B.A. degree and a B.A.Sc. degree in engineering. After prospecting for gold for three years, he went to M.I.T. on a fellowship from the Royal Society of Canada and carried on research in radioactivity and geologic age measurement. He received his Ph.D. in 1940.

In World War II, Dr. Hurley worked with the United States Navy, under the National Defense Research Committee, on anti-submarine warfare and underwater ballistics. A year of research in geophysics at the University of Wisconsin followed, and in 1946 he returned to M.I.T., where he now is Professor of Geology and Executive Officer of the Geology Department. Besides his academic work, he has explored for minerals from the Arctic Circle to the Equator, fulfilled research contracts for the Office of Naval Research and the Atomic Energy Commission, and contributed articles to various technical publications, to *Science* magazine, the *American Journal of Science,* and the *Scientific American.*

Dr. Hurley is keenly interested in the art of teaching and in educational policy. He has served on many national scientific committees and for the last several years on M.I.T.'s principal educational policy committees. It is his conviction that today's educational program needs to emphasize a "broader understanding of our physical environment and a reaffirmation of human values." He takes particular enjoyment in working in the laboratory with students exploring earth history, and he hopes *How Old Is the Earth?* will "start some new explorers in earth science."

Dr. Hurley, his wife, and their three children live in Lexington, Massachusetts.

HOW OLD IS THE EARTH?

Patrick M. Hurley

GREENWOOD PRESS, PUBLISHERS
WESTPORT, CONNECTICUT

Library of Congress Cataloging in Publication Data

Hurley, Patrick M   1912-
    How old is the earth?

    Reprint of the ed. published by Anchor Books,
Garden City, N.Y., which was issued as no. S5
of Science study series.
    Bibliography: p.
    Includes index.
    1. Earth--Age. 2. Geological time. I. Title.
II. Series: Science study series ; S5.
[QE508.H87 1979]   525   78-25843
ISBN 0-313-20776-3

First published in 1959 by Doubleday & Company, Inc.,
New York

Figures 21, 22, 23, 25 have been adapted from ones appearing
originally in the *Scientific American.*

Illustrations by R. Paul Larkin
Typography by Edward Gorey

Reprinted with the permission of Doubleday & Company Inc.

Reprinted in 1979 by Greenwood Press, Inc.
51 Riverside Avenue, Westport, CT 06880

Printed in the United States of America

10 9 8 7 6 5 4 3 2 1

The Science Study Series

The Science Study Series offers to students and to the general public the writing of distinguished authors on the most stirring and fundamental topics of physics, from the smallest known particles to the whole universe. Some of the books tell of the role of physics in the world of man, his technology and civilization. Others are biographical in nature, telling the fascinating stories of the great discoverers and their discoveries. All the authors have been selected both for expertness in the fields they discuss and for ability to communicate their special knowledge and their own views in an interesting way. The primary purpose of these books is to provide a survey of physics within the grasp of the young student or the layman. Many of the books, it is hoped, will encourage the reader to make his own investigations of natural phenomena.

These books are published as part of a fresh approach to the teaching and study of physics. At the Massachusetts Institute of Technology during

1956 a group of physicists, high school teachers, journalists, apparatus designers, film producers, and other specialists organized the Physical Science Study Committee, now operating as a part of Educational Services Incorporated, Watertown, Massachusetts. They pooled their knowledge and experience toward the design and creation of aids to the learning of physics. Initially their effort was supported by the National Science Foundation, which has continued to aid the program. The Ford Foundation, the Fund for the Advancement of Education, and the Alfred P. Sloan Foundation have also given support. The Committee is creating a textbook, an extensive film series, a laboratory guide, especially designed apparatus, and a teacher's source book for a new integrated secondary school physics program which is undergoing continuous evaluation with secondary school teachers.

The Series is guided by the Board of Editors of the Physical Science Study Committee, consisting of Paul F. Brandwein, the Conservation Foundation and Harcourt, Brace and Company; John H. Durston, Educational Services Incorporated; Francis L. Friedman, Massachusetts Institute of Technology; Samuel A. Goudsmit, Brookhaven National Laboratory; Bruce F. Kingsbury, Educational Services Incorporated; Philippe LeCorbeiller, Harvard University; Gerard Piel, *Scientific American;* and Herbert S. Zim, Simon and Schuster, Inc.

8

CONTENTS

9

CONTENTS

Introduction

"Before the hills in order stood,
Or earth received her frame . . ."
ISAAC WATTS (1674–1748)

Ever since man has had the ability to reason, he
has speculated upon the universe about him. The
nature of the earth and stars and their origins have
been described ten thousand times, in different
tongues, in prehistory and history. Each of these
descriptions has rested on some evidence, some
witnessed phenomenon. But only since the seven-
teenth-century renaissance in science has there
been a conscientious effort to seek explanations for
all phenomena within the bounds of a minimum
number of physical laws, to eliminate superstition
and dogma from the realm of science.

This extension of the disciplines of reasoning to
an explanation of our earth, the solar system, the
galaxies, and the universe has led to present-day

11

hypotheses that only new evidence—more powerful applications of physical principles or a more comprehensive fitting of them to the myriad observations of fact—can challenge. Any theory will be modified as time goes by. But lest this present history of the earth be dismissed as visionary, let us remember that to be accepted, even temporarily, a theory must have been stripped of fancy and defended before the scientific world of the day.

The development of scientific thought on the origin of the earth is a particularly constructive record for the student of science: it illustrates the struggle to bring together diverse and conflicting hypotheses and observations and shows how widely opinion can swing in a few years. Indeed, within the lifetime of the youngest reader of this book new knowledge of the atomic nucleus has altered radically our picture of the universe.

The measurement of time by study of the continuous breakdown of radioactive elements has had great impact on science and philosophy. We have learned that the naturally occurring radioactive elements are constantly decreasing in abundance, and this phenomenon forces upon us a new realization. It demands a creation of these elements, and therefore probably of all elements, at some definite time in the not-too-distant past. The elements of the world we live in definitely were not in existence forever; therefore, neither was this earth, nor this solar system, nor our galaxy of stars.

What was the nature of this creation? When did it occur, and how did we reach our present state of being? These are questions for the mature intellect. Studying these processes does not diminish their grandeur. The more clearly we see it, the genesis of this part of the universe, from the birth of the stars to the evolution of the human mind, increases in its inspiring majesty.

Our earth has had a dynamic history, not only in the development of human forms but also in the great changes that have brought to a cold, desolate waste like the moon our present hospitable climates, the lands and seas, the wealth of minerals and fuels, the mountains and plains. What great forces generated these sweeping changes? Mountains are short-lived objects on the earth's surface. Unlike people, they are largest when youngest. Under the eroding actions of ice and rain, they mature to low, rounded forms, and in a time that is short in the earth's history they wear to flat plains, leaving no evidence of their existence except in the rocks and structures of their roots. But as mountains have worn away, other geologic features have appeared. What is the source of the vast energy that develops belts of volcanoes pouring forth molten lava and shaking the ground with earthquakes, that lifts continents above the ocean floor and makes the earth's surface buckle and wrench? What energy concentrates copper and gold into deposits that glitter under the miner's lamp like a jewelry store?

Now we recognize the answer. It is the energy of the radioactive breakdown of nuclei of atoms like uranium and thorium. This occurs in small amounts in all the ground we stand on. The energy released is transformed into heat which, as it has for billions of years, flows to the surface. When the shift of surface material, such as deposition of sediment at the edge of a continent, blocks this flow, the temperature beneath builds up to the point where rocks melt, and the crust becomes weak enough to buckle.

Not only has radioactivity supplied most of the energy for the earth's great geologic events; it also measures the time at which these events have occurred. As we shall see in later chapters, each grain of sand, each minute crystal in the rocks about us is a tiny clock, ticking off the years since it was formed. It is not always easy to read them, and we need complex instruments to do it, but they are true clocks or chronometers. The story they tell numbers the pages of earth history.

But let us start at the beginning. . . .

CHAPTER I

The Structure of the Earth

The earth is almost a sphere, but the centrifugal force of its rotation causes a bulging at the equator and a slight flattening at the poles. Its radius is about 6400 kilometers (or 3960 miles). We know that the earth is divided into distinct layers, and we believe that it is composed of two main types of material.

If you could see the earth in cross section, you would find a rather sharp division between the *core*, or central part, and the *mantle*, or outer part. This is shown in Fig. 1. As you can see in the diagram, the core itself appears to be divided into inner and outer parts. There are several lines of evidence which lead to the belief that the core is composed 90 per cent of metallic iron and possibly silicon, in the proportion of three parts to two, and 10 per cent of nickel, largely in a fluid state. The inner core appears to have the properties of a solid.

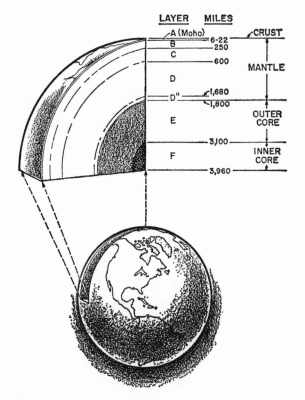

LAYER   MILES

A (Moho)   CRUST
B   6-22
         250
C   600
                  MANTLE
D
D"   1,680
         1,800
                  OUTER
E   CORE
         3,100
                  INNER
F   CORE
         3,960

Fig. 1. The earth's surface is well known to us, but to explain the nature of its interior, we still have only hypotheses, and they conflict. The deepest hole penetrates only about 5 miles. Information about the interior is derived from seismic wave velocities, meteorites, volcanic eruptions, and other measured and observed phenomena.

The existence of a sharp boundary between core and mantle is inferred from observed changes in the direction and speed of shock, or *seismic,* waves passing through the earth. These changes, which can be measured with an instrument called the *seismograph,* indicate differences in the physical properties of the core and mantle. On the evidence of the seismograph, R. D. Oldham of England established in 1906 that the core exists. In 1914, Beno Gutenburg of Germany, later a professor at the California Institute of Technology, showed that the separation between core and mantle occurs about halfway between the surface and the center of the earth. This boundary is called the *Wiechert-Gutenburg discontinuity.* (Ernst Wiechert of Germany contributed to the development of the seismograph.)

The mantle is believed to be composed predominantly of oxygen, magnesium, and silicon (with a little iron) in a ratio of four atoms of oxygen to two atoms of magnesium and one of silicon; all the other elements combined equal only about one tenth the total amount of these four. If a piece of the mantle were brought to the surface, melted, and allowed to cool slowly, it would look like dark green rock, and it would be made up largely of the mineral olivine, magnesium iron silicate. Because of the high pressure and temperatures existing in the mantle (white hot in the lower part), it is not known in which principal mineral forms these elements occur, but it *is* known that they are in the solid state.

17

The outer surface of the mantle has a very shallow skin layer of different composition, known as the *crust*. One of the sharp changes in seismic wave velocity I have mentioned marks the boundary between the crust and the rest of the mantle. A Croatian seismologist, A. Mohorovicic, discovered it in 1909 while studying the seismograph of an earthquake in the Balkans, and it gets its name from him, the *Mohorovicic discontinuity*, or *Moho*. We find the Moho at different depths in different regions but generally at about 35 kilometers (22 miles) under the continents and 10 kilometers under the oceans.

The composition of the crust is quite varied; it ranges from light-colored rocks like granite, rich in silica, alumina, and alkalies, to dark-colored rocks like those of the Hawaiian Islands, rich in iron oxide and magnesia. A thin layer of sediments, which have turned into rocks like shale, sandstone, or limestone or, if buried deep enough, have been baked into more crystalline rocks, covers a large extent of the crust surface. These rocks of sedimentary origin make up only a small fraction of the crust's total mass. The surface of the earth also has masses of water, referred to as the *hydrosphere,* and an envelope of gases, called the *atmosphere*.

The hydrosphere and atmosphere aside, eight elements predominate in the composition of the earth's crust; they make up about 99 per cent of its mass. In terms of numbers of atoms, oxygen accounts for more than 60 per cent of the total.

THE COMMONER ELEMENTS IN THE EARTH'S CRUST
COMPARED WITH AVERAGE STONY (CHONDRITIC) METEORITE

Element	Atomic radius ($\times 10^8$ cm)	Crust		Meteorite	
		Weight (Per cent)	Atom (Per cent)	Volume (Per cent)	Atom (Per cent)
Oxygen	1.32	46.60	62.55	91.97	58.6
Silicon	0.39	27.72	21.22	0.80	16.7
Aluminum	0.57	8.13	6.47	0.77	1.5
Iron	0.82	5.00	1.92	0.68	6.3
Magnesium	0.78	2.09	1.84	0.56	14.9
Calcium	1.06	3.63	1.94	1.48	1.12
Sodium	0.98	2.83	2.64	1.60	.77
Potassium	1.33	2.59	1.42	2.14	.08

(Clark and Washington, Goldschmidt, and Brown and Patterson)

19

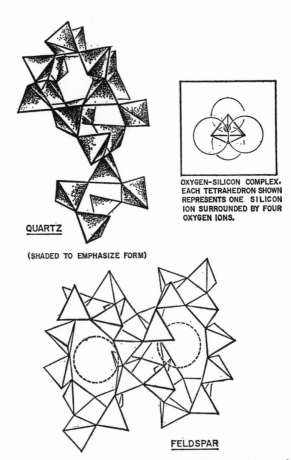

QUARTZ

(SHADED TO EMPHASIZE FORM)

OXYGEN–SILICON COMPLEX.
EACH TETRAHEDRON SHOWN
REPRESENTS ONE SILICON
ION SURROUNDED BY FOUR
OXYGEN IONS.

FELDSPAR

Fig. 2. Oxygen is a component of most rocks and therefore, although commonly thought of as a gas, is one of the principal parts of the earth's crust. The relative proportions of oxygen to the other important elements can be visualized in these schematic draw-

Rocks are *oxygen compounds.* They are composed of such different minerals as quartz, feldspar, and mica, but all these can be visualized as three-dimensional networks of oxygen held together by other atoms. When you consider the relative volumes of the crust elements, it is striking that the hard earth you walk on is for the most part an oxygen platform. This is shown on page 19 in the table of the elements that compose 99 per cent of the crust.

Scale models of the atomic arrangement in the common rock-forming materials show more vividly the predominance of oxygen. In Fig. 2 are models of the crystal structure of quartz and feldspar. The large white spheres are oxygen atoms, the small black ones silicon. Atoms of other elements bind the combinations of oxygen and silicon together in ways that form loosely bonded chains, sheets, or three-dimensional networks. These different forms of crystal structure (Fig. 2) determine how minerals break—in long splinters (hornblend) or in sheets (mica) or in shell-like shapes (quartz). Because different groupings of oxygen with silicon are basic in the structure of most common rock minerals, they are called *silicate minerals.*

ings of quartz and feldspar crystals. Silicon is the next most important element, and most rock-forming minerals are mainly silicon-oxygen compounds known as silicates. (Figure taken from model structures of Dr. Tibor Zoltai, M.I.T.)

How We Have
Learned about the Earth's Interior

Since ancient times earthquakes, volcanoes, and what we now recognize as evidence of evolution have fascinated thinking men. Xenophanes of Colophon (c. 500 B.C.) found sea shells in the mountains. The great Greek philosopher Aristotle (384–322 B.C.) thought winds inside the earth caused earthquakes, and he believed that rain falling on the heated earth "generated" the winds. In his *Metamorphoses* the Roman poet Ovid (43 B.C.–A.D. 17) described many geologic processes. The Renaissance's universal genius, Leonardo da Vinci (1452–1519), insisted that fossils discovered in the hills of northern Italy once had been living creatures. René Descartes (1596–1650), the French mathematician who invented analytic geometry, advanced the hypothesis that an incandescent mass like the sun cooled to form the earth with its crust enclosing a still hot nucleus.

All this was merely speculation. It was not until around 1776 (a convenient date for Americans to remember) that geology became a science. Independently, and more or less simultaneously, A. G. Werner (1750–1817), E. F. von Schlotheim (1764–1832), and William "Strata" Smith (1769–1839) discovered that rock layers can be identified because certain fossils are found in them and not in layers above or below them. This secret revealed, geologists went to work all over the world, and in a little more than a century and a half their

22

successors have built up a great body of knowledge of the physical and chemical nature and the history of the earth's crust. But they cannot investigate the mantle or core of the earth directly and still must infer the composition and physical nature of the interior regions from evidences available at the surface.

An important approach to knowledge of the earth's interior is study of the way shock waves travel through it. This branch of science is known as *seismology* and its main instrument is the seismograph, first developed in Japan in the late nineteenth century by an Englishman, John Milne. Reduced to its essentials, the seismograph is a rigid frame anchored to bedrock; a heavy weight is delicately suspended from the frame. Movements of the frame in relation to the suspended weight, which remains stationary, are recorded with electrical devices. It is sensitive enough to detect minute vibrations of the earth.

Seismological studies of the earth depend largely on earthquakes for sources of seismic waves, because dynamite explosions are not powerful enough. (Hydrogen bombs would work, but they, needless to say, are not generally available.) However, many earthquakes occur, and they have afforded much information. Seismological stations have been established in many countries to record motions of the earth's surface; a single earthquake is recorded simultaneously all around the world. Synchronization of timing devices by radio signal makes it possible to measure precisely the time of arrival of earthquake-generated seismic waves at all the stations.

Two kinds of seismic waves travel from the *focus* of an earthquake: primary (*P*) waves, which

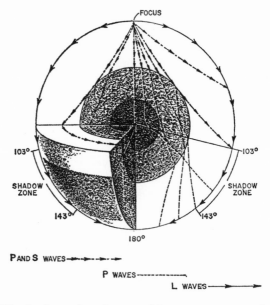

Fig. 3. An earthquake caused by a rupture near the surface of the earth starts the types of waves illustrated in this diagram. Seismic waves travel outward from the earthquake focus as expanding nearly-spherical wave fronts. These are shown diagrammatically in the sketch as "rays," or lines drawn perpendicular to the wave fronts. These are bent by refraction when they encounter changes in physical properties at depth. The core boundary is inferred from the "shadow zone" caused by a sharp change in refraction at this depth.

24

are *compression-expansion* waves like sound waves, and secondary (*S*) waves, which are *shear* waves transverse in motion as are the waves that travel along a stretched string. In addition there are surface waves. The *P* and *S* waves radiate in all directions from the focus, Fig. 3.

Besides indicating the existence of the Wiechert-Gutenberg discontinuity some 1800 miles below the surface, the behavior of *P* and *S* waves gives us data from which to infer something about the properties of core and mantle. It has been found that the waves are refracted (bent upward) in the region of the mantle and reflected downward at the crust. *P* waves propagate through both core and mantle. *S* waves propagate through the mantle but not through the core. From the study of shock waves in liquids and solids, it is known that liquids do not transmit shear or transverse waves like the *S* waves; hence, it is inferred that the core of the earth, or part of the core, is liquid.

Study of shock waves in solids has shown that the velocity of compressional waves (if you stretch a spring and let it go, the pulse traveling up it is a compressional wave) depends upon the density of the solid and its elastic constants, which are numbers that help to describe its reaction to compressing, twisting, and shearing forces. The upward refraction of compressional waves in the mantle indicates the rate at which the combination of density and elastic constants changes with depth.

In the first forty years of this century "travel time" tables for seismic waves were compiled and laboriously corrected. With them geophysicists can

25

compute wave velocities in the earth's interior regions. Changes in wave velocity provide significant information about structure. (A change of P wave velocity from about 7 kilometers a second in the crust to about 8 kilometers a second in the mantle indicated the existence of the Mohorovicic discontinuity at the base of the crust.) By studying velocity changes in relation to the densities and elastic constants of solids and co-ordinating this information with what else is known about the earth and meteorites, scientists have identified seven distinct shells in the earth and come to the conclusion that the core is composed largely of iron and the mantle largely of a rocklike material.

How do meteorites come into the picture? They are small objects from within the solar system that sometimes fall to earth. If these bodies were formed in a manner similar to the formation of the earth, it is possible that they were composed of similar material. Their composition, therefore, may give us a clue to the nature of the interior of our own planet. Two principal classes of meteorites occur: the "iron" meteorites and the "stony" meteorites. The iron meteorites are largely native iron—that is, iron not allied with other elements—and have nickel and other metallic components in minor amounts. The stony meteorites are like dark green volcanic rocks but are richer in magnesium. (See the table on page 19.)

The study of meteorites, then, leads to the assumption that the two main types of material in the earth are metallic iron and silicates (or oxides), separated as they have been in the meteorites, and it is possible to relate this hypothesis to the physical characteristics of the earth—its weight, moment of inertia, and distribution of density. Briefly, the reasoning goes like this.

The gravitational field at the surface of the earth can be measured, and with this figure the total weight of the earth is computed as 5.98×10^{27} grams. From a measurement of the size of the earth, its average density then is found to be 5.515 grams per cubic centimeter. Now, it long has been known that the density of rocks that have risen to the surface in the form of volcanic lava and intrusive masses does not exceed about 3.3 grams per cubic centimeter. Therefore, to make up the average, density must increase toward the center. This would be expected as a consequence of pressure alone, because pressure would tend to compress the material at great depth. But theoretical considerations of the rate of change of density with pressure in materials of the stony meteorites' composition call for a different composition. If we are to account for the necessary density of the central part, a heavier material is needed. This falls in line with the hypothesis that the central part is iron.

When we take into consideration the earth's shape, a theoretical approach to the problem of density distribution is possible. You will remember that because the earth rotates on its axis the radius at the equator is greater than at the poles. Now, if we assume that the earth has responded as a perfect liquid to the combination of centrifugal and gravitational forces, and if we assume different density distributions within the mass as a whole, we can calculate the difference between the radii at equator and poles. If most of the heavier materials were concentrated in the earth's outer shells, the difference in radii would be greater than it is with most of the heavier materials concentrated at the center. Therefore, the actual measured difference of the radii at equator and poles does provide a limit to the nature of the density distribution, but it does not specify exactly what that distribution will be within the limit.

Fig. 4 summarizes the present ideas regarding density distribution in the earth, as calculated by the New Zealand mathematician and seismologist K. E. Bullen. The density distribution, coupled with seismic wave velocity data and the relative abundance of the chemical elements found in meteorites, is the most important factor in our estimates of the earth's interior. As we have seen, study of wave velocity at different depths in the mantle has given information about the existence of the discontinuities and about the relationship between elasticity and density at different depths. In later chapters we shall study the part radioactivity has played in the composition of the upper mantle region rich in such elements as silicon, aluminum, sodium, potassium, uranium, thorium, and others, and in the composition of the crust, the oceans, and the atmosphere.

The Earth's Crust

The crust has two principal topographical features: the *continents* and the *ocean basins*. The

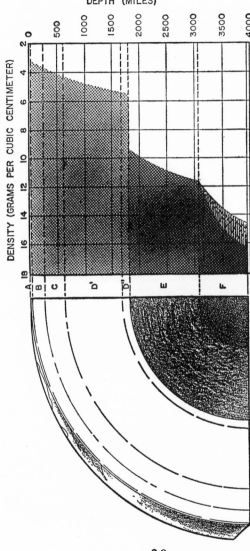

Fig. 4. Density of the earth's interior probably varies in the manner indicated by this graph, which has been developed largely from measured changes in seismic velocity with depth and a knowledge of the effect rotation has on the earth's shape.

29

basins lie approximately 5 kilometers below the
surface of the water and in general form a rather
monotonous plain. There are some areas of intermediate depth, but there definitely is not a continuous and uniform gradation in elevation between
the ocean depths and the land surface. Similarly,
the continental areas lie slightly above sea level
with a relatively small portion rising to high altitude. The graph in Fig. 5 shows the strikingly small

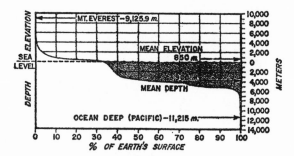

*Fig. 5. The major part of the earth's surface is at a
fairly uniform level of about 5 km below sea level.
Superimposed on this is another large area slightly
above sea level, representing the continents. Then, a
small proportion of the total area lies above, below,
and in between these two dominant levels, representing
the high mountains, ocean deeps, and edges of continents, respectively. There is something fundamental
about these two dominant levels.*

area of the surface that rises much above sea level
or lies above or below the average 5-kilometer
depth of the ocean basins.

There must be something fundamental in the

dynamic history of the earth's surface that has caused continents to exist. Since the ocean area is four times the continental area and since the continents, as we shall see, have undergone considerable change, geologists believe that the ocean basins represent the more or less original, or primordial, surface of the earth. The continents are something different that must be explained.

The most commonly accepted hypothesis to explain the stability of the continents and the ocean basins on the earth's surface had its beginnings in the observations and conclusions of three American geologists—James Hall (1811–98), John Wesley Powell (1834–1902), and Clarence Dutton (1841–1912)—and has since been fortified by gravitational measurements all over the world.

Hall, who published a monumental thirteen-volume work on paleontology, the study of past geologic periods by means of their fossil remains, was the first to theorize that the enormous weight of tremendous accumulation of sediments along the shores of ocean basins caused depressions in the crust of the earth and that these depressions always preceded the process of mountain making. Mountains, he said, were not a product of upheavals or convulsions of the earth but were part of continental movements.

Powell led the first boating expedition through the Grand Canyon of the Colorado River and made a study of the action of water on the rocks of the Colorado Plateau geologic district. He concluded that the process of *downwearing* was irresistible, that in time rain and rivers would wear any

31

surface down to a *base level,* which would be the level of the ocean surface, but no farther. The higher the district, he concluded, the faster the wearing down.

Dutton, a United States Army officer, was a protégé of Powell and, like him, a student of the geology of the Grand Canyon district. He reasoned that the earth would have a perfectly smooth surface if it were composed of homogeneous material, but since the surface is not smooth, some areas of the crust must be made of lighter materials than others. In the less dense areas—mountains and plateaus—the surface would tend to bulge; in denser areas the surface would be depressed in basins which might become filled with sediments. It was plain to be seen in the mountain ranges of the West, he said, that the vast platforms on which the mountains rest had continued to rise as fast as erosion degraded the mountains themselves.

Dutton summarized this state of balance as a crustal equilibrium resulting from the force of gravity, and he coined the name *isostasy* (from Greek words meaning "standing still") for it. Gravity measurements in the Himalayas and along the Atlantic and Pacific coastlines were in accord with his theory. Plumb lines are deflected from the vertical toward the denser masses of the crust. As illustrated in Fig. 6, the surface topography of the earth results from the "floating" of materials of lower density and varying thickness in a material of greater density beneath.

Although the material of the mantle is believed to be solid, the stresses under the weight of con-

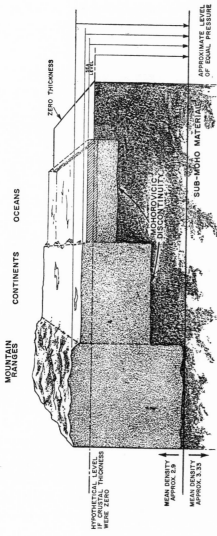

Fig. 6. Isostasy is a well-established concept which holds that the continental masses are higher than the ocean basins because they are lighter. The solid rock beneath yields slowly to the great stresses involved until heights of land are buoyed up above the general level of the earth's surface, with the height dependent upon the thickness or relative density of the lighter material.

MOUNTAIN RANGES   CONTINENTS   OCEANS

ZERO THICKNESS

SEA LEVEL

MOHOROVICIC DISCONTINUITY

SUB-MOHO MATERIAL

APPROXIMATE LEVEL OF EQUAL PRESSURE

HYPOTHETICAL LEVEL IF CRUSTAL THICKNESS WERE ZERO

MEAN DENSITY APPROX. 2.9

MEAN DENSITY APPROX. 3.33

tinental masses are so great that the mantle, behaving as a viscous fluid would, almost surely must be deformed.

Rocks that make up the continental crust are almost certainly lighter, or less dense, than the material of the mantle underneath. Therefore, like a great layer of shelf ice floating in a sea, a continent is visualized as a layer of lighter rocks floating in gravitational equilibrium in a medium which, although it is solid rock itself, will yield to the stresses involved, given enough time. Like ice, whose height above the water level is dependent upon the depth of the ice below the water level, so apparently the crustal materials float on the universal mantle material beneath.

In other words, when the thickness of the lighter crustal layer is small, it floats at a lower elevation than when it is large. Therefore, the crustal-type rocks in the ocean basins are believed to be much thinner than the layer in the continental areas. This is borne out by measurements of the depth of the Mohorovicic discontinuity, which show a 35-kilometer average depth under the continents and a 10-kilometer depth below sea level under the ocean basins.

A little arithmetic will help to clarify this point. For example, if the average density of the material of the mantle below the crust is 3.33 grams per cubic centimeter, and the average density of the continental crust above the Mohorovicic discontinuity is 2.90 grams per cubic centimeter, the density difference of 0.43 is approximately $\frac{1}{8}$ of that of the medium below the discontinuity. This means

34

that a displaced volume will have a buoyancy that will carry approximately ⅛ of its volume as additional volume above the surface of the denser medium. In other words, if the continent is floating at an elevation of 4.5 kilometers above a hypothetical mantle surface, this elevation should be supported by a submerged part that is $4.5 \times 7 = 31.5$ kilometers below this surface or about 36 kilometers below sea level. This is approximately what is observed if the Mohorovicic discontinuity is in truth the base of this submerged part.

Similarly, in the column beneath the oceans the "crust" would consist of, say, 5 kilometers of water of density 1 and 5 kilometers of rock of density 2.8, or an average density of approximately 1.9, so that the density difference in this case would be 1.43. This is more than 3 times the density difference found for the column of continental crust; so the thickness would be expected to be ⅓ as much. This conclusion is again supported by the measured depth of the Mohorovicic discontinuity under the oceans—namely, 10 kilometers. If these features of topography were truly floating in gravitational equilibrium in this manner, the measurement of gravity over the surface of the earth (corrected for altitude and latitude) should all be the same. This is found to be in large part true, except for zones of disturbance in which active mountain-building processes have gone on in the recent past so that isostatic equilibrium no longer exists.

Commonly located along the margins of continents, but not always, these zones, which are out of equilibrium and distinctly different from the stable oceanic and continental areas, are of great interest in geology because they tend to throw light on the possible origin and development of mountains and continents in the past history of the earth.

Mountains and Volcanoes

We can see that there is a need for a large expenditure of energy in the formation of these belts of mountains. If some dynamic process develops a region that is out of gravitational equilibrium, there will be continuing stresses and yielding in the direction of restoring the equilibrium. Consider again the example of the mass of ice floating on water: If the ice above the water level is melted by warm winds and sun, more ice will be fed upward by the entire mass rising as the load is reduced, until finally the ice near the bottom of the mass, originally deep under water, will be exposed at the surface.

At least some mountain belts, like the Sierra Nevada range in California, appear to owe their elevation to deep roots of lighter material. As the mountains are eroded by rain and ice and chemical decomposition, the root would be expected to be out of gravitational equilibrium and rise like the ice mass. But the mountains can be worn down only to sea level, which is several kilometers higher in elevation than the ocean floor; so the root would stop rising, as we have seen, at a depth of about 35 kilometers. Because of the presence of the ocean, therefore, it is possible to have permanently stable masses like continents on the earth's surface, and the explanation of the Mohorovicic discontinuity becomes evident.

Thus we see that whenever a mass of lighter material develops, it will be unstable until worn down to sea level and until it reaches gravitational balance. After that time it should be a permanent fixture on the earth's surface, unless processes are at work other than those considered here. This fact is borne out by the observation that although large areas of the continents are very old, they have not disappeared or sunk appreciably but have remained almost at sea level since they were first eroded to this level and the base of their roots had risen to the 35-kilometer mark.

How are these lighter masses formed? This question is strongly debated among geologists, but one hypothesis is that when a part of the upper mantle melts, as a result of insufficient removal of the heat from radioactivity, a separation of the elements occurs. The elements that form stable minerals at higher temperatures and pressures remain at depth, while the elements that become semi-liquid at these temperatures and pressures rise toward the surface. Not only is there some separation of these semi-liquid components toward the surface, but they also recrystallize near the surface into mineral forms which are less dense than those containing the elements originally at depth before the melting happened.

For example, sodium, calcium, aluminum, and silicon combine with oxygen to form the relatively light feldspar structure in the crustal zone near the surface, but undoubtedly they were held in more dense crystal structures before they were released from their earlier mantle location.

Let us say, therefore, that one way in which the lighter continental masses can come into existence is by a heating of the upper mantle to a temperature of at least partial fusion, with a resultant upward separation of certain elements and recrystallization into less dense minerals. We have already noted how, by the build-up of thick masses of sediment at the margins of continents, it is possible for the flow of heat to be impeded in these areas so that melting at depth can occur. And we know too that the greatest mountain belts have occurred along the former margins of continental areas. There is evidence in some parts of the earth that the continents are growing in size outward, by adding stable new belts at their periphery, with the material coming in part as sediments from the erosion of the continent, but mostly from the mantle below.

Of course, not only are there many variations on these ideas, but also many different hypotheses; and it is probable that several different processes have been active in different areas and at different times in earth history. Nevertheless, all possible processes have the common requirement of energy. So let us examine the sources of energy and their limits in more detail.

CHAPTER II

Radioactivity

At this instant the reader, if he weighs about 170 pounds, is being bombarded internally by approximately 700,000 damaging bullets per minute. These bullets are sufficiently powerful to break chemical bonds, modify molecular structures, and even destroy cells. The reader is not conscious of this constant disruptive bombardment because his body has the power to repair the damage much more rapidly than it is caused. But though the truth is not completely understood, it is possible that these natural radiations may be a potent factor in the life span of a living organism or in the evolutionary process whereby organisms change slowly throughout time into different forms.

To measure radiation dosage, science commonly uses a unit called the *rad,* which represents the quantity of energy dissipated per gram of matter, or:

1 rad $= 100$ ergs $= 2.4 \times 10^{-6}$ calorie per gram.

(A unit called the *rem* also is used; it represents an amount of energy absorbed, with correction for the relative biological effects of the different kinds of radiation.)

If we are to gain a perspective on the relative importance of the possibly injurious sources of radiation—the first that springs to mind is, of course, fallout from tests of nuclear devices—we shall find it interesting to compare the doses that human bones receive from *all* sources. The table opposite, summarized in 1958 by R. A. Dudley, gives instructive estimates, which no doubt will be revised upward as our studies continue.

From this table you can see that the doses received from natural radionuclides of the uranium and thorium series and potassium, both in the body and the outside environment, and from cosmic rays, which always have bombarded inhabitants of the earth, have been roughly equaled by the new sources of radiation, such as medical X-rays and fallout. Furthermore, commercial power generation from fission is expected to reach 30,000 megawatts by 1975, and it will produce fission products at a rate equal to 300 megatons of bombs per year. This greatly exceeds weapon-testing rates; it will be important to prevent even a small proportion of these fission products from escaping.

(The isotopes in the first line of the table are called *radionuclides* because they are radioactive and because they are species of atoms characterized by the constitution of their *nuclei;* that is, by the numbers of protons and neutrons their nuclei

40

contain. The symbol K stands for potassium, Ra for radium, and Pb for lead.)

ESTIMATED SKELETAL DOSE RATES OF MAN

Source of Radiation	*Avg. Skeletal Dose Rate millirad/year*	Comment
Naturally radioactive isotopes present in the bone (K^{40}, Ra^{226}, Ra^{228}, Pb^{210})	34	Average value only. Variations within a factor of 3 to 5.
Strontium 90 and Caesium 137 from fallout	3 (1957, or 12 for 1975)	From past explosions only.
Cosmic rays	30	Value near sea level at higher latitudes.
Naturally radioactive isotopes present in environment	45	Typical value for sedimentary rock area; allowing for building materials.
Gamma rays from fallout	0.5	In U.S., 1951–56
Medical X-rays	75	Value very variable and inaccurately known.
Total	200	

To safeguard individuals working with radioactive materials, the Atomic Energy Commission has proposed regulations to limit the basic permissible tissue dose for exposure to any ionizing radiation. For an indefinite period of years the dosage for internal organs could be 100–300 millirads per week. Thus, the natural and man-made radiations still amount to only $\frac{1}{50}$ of the AEC's specified limits for the safety of workers.

Radioactive Fallout

Localized concentrations of short-lived radioactive isotopes formed in bomb tests received considerable publicity a few years ago. For instance, after the H-bomb tests in the Bikini Atolls in March 1954 considerable amounts of radioactive zinc 65 and other isotopes were found in tuna fish arriving in Japanese ports. Since the Japanese people derive 90 per cent of their protein diet from sea products, the hazard was considered so serious that the products were not easily sold on the market. In the month following the bomb tests the market prices in the principal fishing port of Misaki dropped to less than one half. Contaminated fish were caught at great distances from Bikini, even in waters surrounding Japan.

The early scares over such fallout instances now have given way to careful estimates of the long-range effects of nuclear fallout and contaminations, and the estimates are based on many measurements and sober calculations. Most of the present-day concern is with the level of concentration of a few long-lived isotopes, of which much the most important is strontium 90. Since this isotope, which has a 28-year half-life, is gradually brought to earth by rain over a period of several years following nuclear explosions, its concentration will gradually increase. Cows eating grass take it up from the ground, and it eventually gets into the human system through the consumption of milk and dairy products. To a small degree it supplants calcium in the bones.

The 1958 report of the Advisory Committee on Biology and Medicine, of the National Academy of Sciences, came to the conclusion that the population of the United States should not, as a genetic safeguard, receive a radiation dose exceeding an upper limit of 10r in thirty years. At this writing (1959), United States residents, it is estimated, have on the average been receiving from fallout over the past five years a dose which, if weapons testing were continued at the same rate, is estimated to produce a total thirty-year dose of about 0.1r, or $\frac{1}{100}$ of the estimated permissible limit. The average world-wide thirty-year dose was estimated to be considerably lower. Thus, the report suggested that there was not yet any cause for alarm, but in view of the adverse repercussions caused by the testing of nuclear weapons, it recommended that tests be held to a minimum consistent with scientific and military requirements.

In short, therefore, radioactivity is all around us and we cannot escape it. The level of background radioactivity has been part of the environment of living creatures, and the evolution of life on the earth has become thoroughly adjusted to it. Man-made concentrations or disturbances in this realm could possibly cause harmful concentrations, but with the correct scientific vigilance it should be possible to keep control of this danger, unless the matter becomes one of malign intent in war. Let us, therefore, not fear radioactivity, but use it to our advantage.

Radiation and Particles

What is the nature of these common radiations and fast-moving particles? Where do they come from? How does the physicist produce them and use them? Before we go on to examine, in subsequent chapters, radioactivity's role in measuring the age of the earth and the part it has played in shaping our destiny, we should have at least a general understanding of some fundamentals.

First we must have a glossary of terms. In the strange new world of nuclear physics the scientist finds himself able to observe phenomena and measure quantities and effects that he does not understand fully. Names are given to these observed entities while they are being studied, but generally these names do not last very long because the entity is found to be made up of more basic entities; the name becomes superfluous. It is highly instructive for the student of physics to study the historical development of every phase of the science. We shall do this briefly in later pages, but first let us have a list of the present names given to the quantities in our subject and omit the confusing earlier terminology. For example, terms such as canal rays, cathode rays, and positive rays have historical interest to the physicist, but actually are obsolete.

Therefore, keeping in mind that giving something a name merely isolates one or a few of its properties and does not describe it fully, we shall list some commonly used terms and their principal characteristics.

The *nucleus* of an atom is currently described, in terms of two particles, *neutrons* and *protons*, which exist in roughly similar proportions in most of the elements. As the name implies, the neutron is an electrically neutral particle, while the proton has a positive electric charge of 4.8029×10^{-10} electrostatic units. Because energy is required to develop an electric charge, the bringing together of a number of protons will increase the total charge density, or energy, in a nucleus. As a result, the nucleus would tend to fly apart if it were not for some binding force, little understood, that keeps the particles together. The particles are held so close by these short-range binding forces that the density of the nucleus is about 2×10^{14} grams per cubic centimeter.

The difference between an atom of one element and the atom of another element is the difference in total positive charge of the nucleus, or the number of protons in it. The total number of neutrons plus protons in a nucleus is referred to as the *mass number* A. The total number of protons or positive charges in the nucleus is known as the *atomic number* Z. For example, uranium 238, which is the largest naturally occurring nucleus, contains 92 protons and 146 neutrons. A common notation that defines every individual atomic species has the atomic number as a subscript before the symbol for the element and the mass number as a superscript following it; for instance, $_{92}U^{238}$ means that this atomic nucleus has 92 protons and $238 - 92 = 146$ neutrons.

Atoms that have the same number of protons but different numbers of neutrons are still almost chemically identical and are called by the same element name. They are referred to as *isotopes* of the element. For example, oxygen has three naturally occurring isotopes, $_8O^{16}$, $_8O^{17}$ and $_8O^{18}$.

Like the cocking mechanism on a spring-loaded gun, the binding forces that hold the protons and neutrons together in a nucleus must act only at very short range. Thus, when a nucleus is split by some means, the fragments fly apart with great velocity. When too many protons are packed together with neutrons and the binding forces are close to their limit of being able to hold the assemblage together, the normal vibrations within the structure may occasionally exceed the limit of a bond, and a part of the nucleus flies off spontaneously. This is referred to as *radioactivity,* or the radioactive *breakdown* of the nucleus.

In the simple loss of a fragment of the nucleus one of the most usual particles emitted is always made up of two neutrons and two protons. This is the same as the nucleus of the helium atom and is known as an *alpha particle*. The loss of an alpha particle usually leaves the nucleus in an *excited state,* and it does not settle down to a stable *ground state* until it has further emitted one or more *gamma rays.* A gamma ray, like a single photon of light, has properties that make it behave like electromagnetic radiation in some circumstances and a small bundle of energy (equivalent to a particle of matter) in other circumstances.

Another way in which an unstable nucleus can acquire greater stability is by a *transformation* in which the nucleus changes its charge by one and emits an *electron,* together with a *neutrino.* The high-speed electron emitted radioactively from a nucleus is known as a *beta particle.*

In an alpha transformation the mass of the alpha particle plus the mass of the remaining nucleus does not quite equal the mass of the original nucleus. The difference in total mass before and after the *event* represents the mass-equivalent of the energy expended in the alpha particle and gamma rays. The energy and the mass differences are related by the Einstein equation $E = mc^2$. The energy E is given as the kinetic energy of the alpha particle and recoiling parent nucleus and as the radiation energy of the gamma rays; c is the velocity of light, equal to about 3×10^{10} centimeters per second; m is the difference between the mass of the nucleus before the alpha particle was emitted and the mass of the nucleus plus alpha particle afterward. In the beta transformation the energy of the beta particle plus concomitant measurable radiation is not sufficient in terms of mass equivalent of energy to make up the difference in mass before and after the transformation. As a result,

47

the neutrino has been postulated by the Swiss physicist Wolfgang Pauli to make up the difference. The existence of the neutrino has been hard to prove, but in recent experiments it has been detected.

Finally, a nucleus may occasionally undergo another kind of transformation in which it captures one of its associated electrons and emits a neutrino, changing the atom's charge by one. This is referred to as *K-electron capture*. This transition is important in geology, because it provides the basis for the argon-potassium method of measuring geologic time, which we shall describe later in the book.

To summarize, we see that there are alpha particles, beta particles, and gamma rays involved in the radioactivity about us. The alpha particles are heavy, being made up of two neutrons and two protons, and although very energetic, travel only about ten centimeters in air, or the width of a pin point in solid matter. The beta particles are electrons, which travel a little farther than alpha particles in solid matter, but are quickly stopped by their interaction with the surroundings. The gamma rays are like unusually powerful X-rays and will penetrate matter for a distance of several centimeters, but are largely absorbed in the first few centimeters traversed.

With these few brief terms in mind let us now return to the interaction of these radiations and particles with matter.

Absorption of Gamma Rays

Almost everyone has become familiar with the fact that gamma radiations can be detected with portable devices. The prospector roams the hills with a portable Geiger counter or scintillation counter; workers in any area with radiation hazard wear film badges to indicate their exposure to radiation; civil-defense organizations instruct radiation monitoring teams in the use of portable detection instruments.

The range of gamma rays from a radioactive source material depends upon the amount of matter traversed. It is not necessary to have a lead shield about a highly radioactive source, if the source can be kept where people do not get close to it. In other words, radiation intensity is reduced by distance, as well as by shielding materials.

Gamma rays are absorbed by matter. By matter we mean atoms of different kinds in gaseous, liquid, or solid state. The volume of an atom is made up almost entirely of empty space traversed by electrons. The interaction of gamma rays with matter is almost entirely their interaction with electrons. The mass of the electrons is negligible compared to that of the very dense nuclei, but these nuclei make up only a very small part of the field of view. As the total number of electrons usually is equal to the total number of protons in the nuclei, and the number of these is almost proportional to the total mass of the nuclei, the density of electrons in any material is almost proportional

49

to the mass of that material measured by its weight per cubic centimeter.

Thus, lead having a little more than ten times the density of water, gamma rays will be absorbed more than ten times more effectively by lead than by water.

There are two principal ways in which gamma rays interact with electrons.

First, a gamma ray may strike an electron and bounce off it at an angle, imparting a velocity to the electron in the other direction. The kinetic energy given to the electron is taken out of the gamma ray, so that the ray continues on its path with reduced energy. The term "bounce" is a strange one to use for something with the properties of electromagnetic radiation, but when you remember that the electron is a charged entity which acts as either a particle or a wave, depending upon the way it is observed, and that the gamma ray is a coupled electric and magnetic field occupying an extremely small and impossible-to-imagine cross-sectional area in its trajectory through space, the phenomenon cannot be anything but strange in the first place. Let it be said simply that the gamma ray loses some of its energy, changes its direction, and knocks an electron out of its atomic orbit with a velocity that will damage surrounding electronic structures. This effect is known as *Compton scattering*. It is illustrated schematically in Fig. 7. The American physicist A. H. Compton, working with X-rays, discovered it in 1923.

Second, there is an interaction known as the

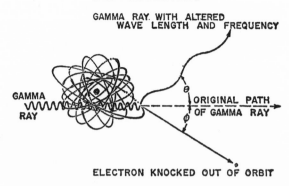

GAMMA RAY WITH ALTERED
WAVE LENGTH AND FREQUENCY

GAMMA
RAY

θ

ORIGINAL PATH
OF GAMMA RAY

ϕ

ELECTRON KNOCKED OUT OF ORBIT

Fig. 7. Gamma rays can dislodge the outer electrons of atoms, losing some of their energy and giving enough to those electrons for them to affect other atoms. This schematic representation of Compton scattering is not a true picture of the atom and gamma rays but does help us to discuss the phenomenon.

"photoelectric effect." This effect happens with photons of light as well as photons of gamma radiation. (A photon is a *quantum,* or discrete "packet," of radiation.) When a gamma ray has lost most of its energy through Compton scattering, it has a high probability of losing the rest of its energy in complete absorption in an atomic electron. Then the gamma ray simply disappears and its entire energy is imparted to the electron.

The sum of these two effects is that gamma rays lose their energy until they disappear, the energy being given up to one or many electrons, which in turn lose their energy by striking or interacting with other electrons until, finally, the energy has been changed into a general excitation of the

51

atoms of the matter traversed, and is now in the form of heat. Sometimes a small proportion of the energy will remain trapped; locally some atoms are in a higher energy state, which gives materials such properties as *thermoluminescence*.

A simple experiment demonstrating thermoluminescence is easy to do. Place some broken fragments of the mineral fluorite (many other minerals will show the same effect) in a frying pan and heat them in a dark room. When the grains have reached a certain temperature, they will suddenly glow for an instant with a dull light. This light is given off as electrons fall back into positions from which they have been driven by gamma radiation over a long period of time. The gamma radiation comes from radioactive impurities in the fluorite.

We see, therefore, that the attenuation of gamma rays is dependent upon two laws. First, like light, gamma rays decrease in intensity with the square of the distance from a point source. Purely a geometrical attenuation, this can be calculated for any shape of source. We consider the source as a large number of small point sources and add the effects of each.

Secondly, the attenuation of gamma rays is one of probability of striking electrons. In any specific case the probability is proportional to the number of electrons present in a unit volume and also to the number of gamma rays present. Let us take, for example, a parallel beam of gamma rays traversing a homogeneous material in which the electron density is essentially constant. This is shown diagrammatically in Fig. 8.

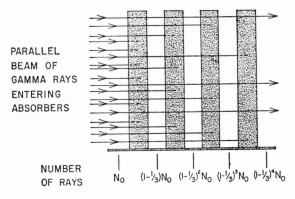

PARALLEL
BEAM OF
GAMMA RAYS
ENTERING
ABSORBERS

NUMBER
OF RAYS

N_0 $(1-\frac{1}{3})N_0$ $(1-\frac{1}{3})^2N_0$ $(1-\frac{1}{3})^3N_0$ $(1-\frac{1}{3})^4N_0$

*Fig. 8. The intensity of a beam of gamma rays pene-
trating a series of equal plates of absorbing material
decreases* exponentially, *as the number of plates (or
total thickness) is increased, as shown above.*

Let us say further that the number of interactions of gamma rays with electrons in the first
centimeter of travel through the material reduces
the number of gamma rays by a certain factor—say,
by one third. In the next centimeter of travel we
start with only two thirds the original amount of
gamma rays; therefore, the probability of interactions in the second centimeter of travel is one
third of two thirds, or two ninths, leaving only
four ninths of the original gamma rays to continue.
In other words, in every successive centimeter of
travel the number of gamma rays emerging from
that centimeter of material will be a constant proportion of the number of gamma rays entering the
material. This proportion will be set by the electron
density of the material.

53

We can readily describe this process in mathematical terms, which, as we shall see, have significance beyond the description of gamma-ray absorption. As so often happens in physics, a law derived for one relationship may apply to a variety of others. In this instance our final equation also applies to the rate of radioactive decay of elements like uranium and thus, as later chapters will disclose, is doubly pertinent to the subject of our book —calculation of the earth's age.

So, if you would care to come with me on a brief side excursion to examine what is commonly known as the "exponential law," I think you will find it fun and pleasantly simple, but if not, you can join us a bit further on, and we shall still arrive at our main destination together.

First we shall call the total distance the gamma rays travel "x" and say that there are "n" number of equal short distances $\triangle x$ in x, just as there are 10 one-inch intervals in 10 inches, 10 being the n, one inch the $\triangle x$, and 10 inches the x ($\triangle$ is the Greek letter *delta*). We shall use N_0 (called "N subscript 0") for the number of gamma rays entering the material, and N for the number of rays left at any point in the material. We shall use another Greek letter μ (*mu*) for the reduction factor, which in the foregoing example we arbitrarily set at one third. Now we are ready to start.

In the short distance $\triangle x$ the material will absorb $\mu\triangle x$ times the number N_0 rays that entered, or $N_0\mu\triangle x$. The number of rays left will be the original total minus those absorbed, or

$N_0 - N_0\mu\triangle x$, which can be expressed as N_0 times $1 - \mu\triangle x$, or $N_0(1 - \mu\triangle x)$.

In traversing the second short distance $\triangle x$ the rays again will be absorbed in the same constant proportion. So, we multiply $N_0 - N_0\mu\triangle x$ by $\mu\triangle x$ to find the number of rays absorbed. We get $N_0\mu\triangle x - N_0\mu^2\triangle x^2$. Subtracting this from the number of rays that *entered* the second short interval, we find the number left at the end of the second interval:

$$N_0 - N_0\mu\triangle x - (N_0\mu\triangle x - N_0\mu^2\triangle x^2)$$
$$= N_0 - N_0\mu\triangle x - N_0\mu\triangle x + N_0\mu^2\triangle x^2$$
$$= N_0 - 2N_0\mu\triangle x + N_0\mu^2\triangle x^2$$
$$= N_0(1 - 2\mu\triangle x + \mu^2\triangle x^2)$$

which is $N_0(1 - \mu\triangle x)^2$, the number of rays left at the end of the second $\triangle x$ distance. We could go on through a third $\triangle x$, getting $N_0(1 - \mu\triangle x)^3$, and a fourth, $N_0(1 - \mu\triangle x)^4$, but it is better to use general terms rather than specific terms. So let us just consider n distances of $\triangle x$. After n intervals of $\triangle x$, the number of gamma rays left will be $N_0(1 - \mu\triangle x)^n$.

Now, as we said earlier, there are $n\triangle x$ distances in the total distance x; so:

$$n\triangle x = x \text{ or } \triangle x = \frac{x}{n}$$

To tidy matters up a bit more, let us substitute $\frac{x}{n}$ for $\triangle x$ in our general expression of the number of remaining gamma rays. Thus, $N_0(1 - \mu\triangle x)^n$ becomes $N_0(1 - \mu\frac{x}{n})^n$, and that brings us to the equation we were seeking, the law stating the number of remaining gamma rays N at any point in the material. We have (see Fig. 8)

$$N = N_0(1 - \mu\frac{x}{n})^n$$

To get an accurate statement of this law, we shall have to make the $\triangle x$ intervals very small. The smaller they become, the larger n, the number of those intervals, must become. (There are more hundredths of an inch in 10 inches than there are tenths of an inch.) The ultimate accuracy comes when we let $\triangle x$ approach zero; then n necessarily approaches infinity.

We shall not prove it here, but the number 2.718, known as e, has the valuable mathematical property that e raised to the xth power equals the limit of $(1 + \frac{x}{n})^n$ when n approaches infinity. We write it

$$e^x = \lim_{n \longrightarrow \infty} (1 + \frac{x}{n})^n$$

If we apply this to our friend $N = N_0(1 - \mu\frac{x}{n})^n$, we get the exponential equation:

$$N = N_0 e^{-\mu x}$$

This equation, as I have said, tells us not only about gamma rays but also about the disintegration of uranium. If we make N_0 the original number of uranium atoms millions of years ago and μ the proportion of atoms disintegrating each second, then N will be the number of atoms of uranium left at the end of any time x. Hence its value in calculating the age of our planet.

And now let us return to our laboratory—the earth and the particles that compose it.

Alpha and Beta Particles, and Heat

When an alpha particle travels through matter, its two protons represent a fast-moving positive charge, which reacts with the electrons in its path, pulling them out of their orbits and imparting an additional velocity to them. The mass of the particle is so great, however, that the interaction of the electrons does not change the direction of the alpha particle to any extent, and it gradually loses energy by accelerating the electrons in the path of its travel. As before, these accelerated electrons gradually transfer their energy to a general excitation of the electronic and atomic structures. This excitation we know as heat.

There will also be some loss of energy from the interaction of the positive charge with the nuclei of the atoms of the matter traversed. But in most cases it is not until nearly at the end of its path that the alpha particle interacts with atomic nuclei. Except for the rare instance when the alpha particle strikes a nucleus in the earlier part of its travel, it usually travels until its velocity drops to a small fraction of its original velocity before it collides or interacts with atomic nuclei. At this time it has sufficient energy left to knock a nucleus out of its position in a crystal structure, and this nucleus

57

gains enough speed from the collision to knock other nuclei out of position until all the energy has been dissipated.

The recoil imparted to the parent nucleus from which the alpha particle came is sufficient to disrupt other atomic nuclei in the material. Because the speed of the recoiling parent nucleus is low compared to the velocity of the alpha particle, it spends most of its energy disrupting other atoms instead of electrons alone. As a result, most of the damage done in the alpha event is done by the recoiling parent nucleus. Several thousand atoms are disrupted at the two ends of the alpha particle's path. Almost all the energy is lost as heat. The damage and excitation of the structure is so intense locally that it is equivalent to a temperature of tens of thousands of degrees centigrade in the disrupted volume. Depending upon the type of material, this damage will remain permanently in the structure, or the structure will gradually return to its crystalline arrangement through the passage of time. Either way very nearly all the energy goes immediately into heat.

The damage caused by a beta particle is rather similar to that of a gamma ray. It will interact with electrons along its path but not disrupt atoms, except by disturbing the bond between atoms and thereby possibly causing chemical rearrangements. It loses energy to the surrounding electronic structure by accelerating the electrons, and again all but an insignificant amount of the energy goes eventually into heating the structure.

So, we see that all radioactive disintegrations

eventually result in a supply of heat to the surrounding material. The amount of heat supplied can be calculated from the known energy of the different particles and radiation. Now, after this brief survey of radioactivity, we are ready to seek the significance of this heat in geologic history, to learn how these radioactive processes have profoundly changed the character of the earth, making it a hospitable place to live on instead of a barren, desolate waste.

CHAPTER III

The Radioactive Earth

Without the heat from radioactivity it is probable that we would have had no atmosphere or oceans. Even if the ocean had existed, no land would have risen above it. Indeed, it is probable that the earth would have had a bare, rocky surface like the moon's, scorched by the sun in daytime and bitter cold at night. You who read this book would never have been born.

But the story of the earth is a story of heat. Throughout earth history large amounts of energy have been continuously expended in mountain building, volcanism, and other activities which have formed the continents, oceans, and atmosphere. Except for the actions of the surface agencies, driven by heat from the sun, the energy comes from the interior of the earth and must have been at one time in the form of heat. To try to explain the occurrence of this thermal energy, we must consider two principal sources: the heat inherited from the formation of the earth and the heat generated in the breakdown of radioactive elements.

There are many arguments in favor of believing that the earth formed at a relatively low temperature. If this is true, a uniform distribution of the radioactive elements that we estimate were contained within the earth would have heated it sufficiently to have caused it to melt or partly melt. It is purely by chance that the sequence of events which we believe followed was such that the bulk of the earth stopped heating up again and remained fairly stable.

These purely chance events are as follows. First, if the mantle of the earth is solid and there is no convection (transfer of heat by movement of fluid material) in it, it must lose heat by the slow process of conduction in the upper regions. If there is convection, from melting or otherwise, the heat can be rapidly transported to the surface. Therefore, the temperature can never get very much above that necessary for melting. In the lower regions heat may be carried out by radiative transfer.

Second, as soon as melting begins, there probably would be a migration of the molten radioactive substances upward because they crytallize at lower temperatures than compounds of magnesium, silicon, and iron. They would be forced upward in the liquid as the solid material settled downward.

If in any region the heat-producing elements have not moved close enough to the surface, the temperature will rise locally to the melting point and a further upward migration will occur. Eventually they will have come close enough to the surface so that there will be no further melting.

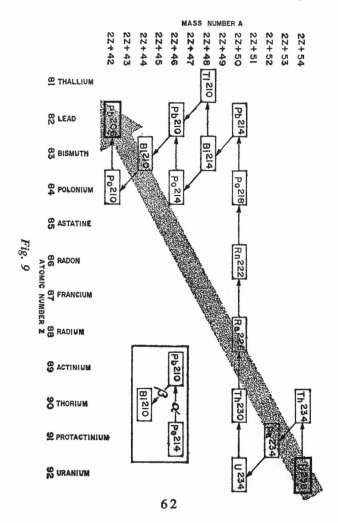

Fig. 9

62

Enough of the heat generated will be lost by conduction to the surface so that a stable solid mantle remains. Gradually, following that time, the radioactive elements will decay slowly, and their heat production will diminish. This would tend to stabilize the mantle so that it is at some temperature below the melting point of its most fusible components. But if the chemistry of the radioactive elements had been otherwise and they had settled into the core, the earth would be continuously melting, losing heat by convection and solidifying again.

The Major Heat-Producing Elements

Let us now examine the amount of heat given off by radioactive elements and estimate what abundance of these elements would cause melting in the mantle. The element uranium breaks down through several stages to form a stable end product, lead (see Fig. 9). As it undergoes successive transformations toward this stable end product, the isotope uranium 238 gives off 8 alpha particles as well as numerous gamma rays and beta particles. Summing up the energies of all these emitted particles and rays, we find that a total of 47.4 Mev (million electron volts) of energy has been ex-

Fig. 9. (*opposite*) *Uranium 238 decays spontaneously to form thorium 234, which in turn breaks down into protactinium 234, and so on until the procession stops at lead 206, which is stable. Some of the transformations are accompanied by alpha particle emission and some by beta particle emission.*

pended for each atom of uranium 238 that breaks down to form an atom of lead 206. Since 1 Mev is equivalent to 3.83×10^{-14} calories, it can be calculated that one gram of uranium in equilibrium with its daughter products is continuously giving off 0.71 calories per year. Similarly it can be calculated that the isotope uranium 235, of which atom bombs are made, is giving off 4.3 calories per gram per year when in equilibrium with its daughter products. Thorium and its series give off 0.20 calories per gram of thorium per year. The only other important heat-producing element is the isotope of potassium, K^{40}. This gives off beta particles and gamma rays at a rate that yields 27×10^{-6} calories per gram of total potassium per year.

Average granite and volcanic rock contain approximately the following amounts of these radioactive elements:

Rock Type	Uranium parts per million	Thorium parts per million	Potassium %
Granitic rocks	4	14	3.5
Dark-colored volcanic rocks	0.6	2	1.0

Thus the radioactive components of the average granite can produce 7 microcalories of heat per gram per year. Other rocks that make up the bulk of the crust produce somewhat less heat than granites, and it is estimated that the average rock in the crust above the Mohorovicic discontinuity probably produces about 2 microcalories of heat per gram per year.

There is a measurable amount of heat continuously flowing to the surface of the earth. Measurements over continental areas have indicated that this amount averages about 1.2 microcalories per square centimeter per second. The amount of heat flowing in oceanic areas has been difficult to measure, but several measurements have been made. This is done by dropping a probe from a ship so it penetrates the mud on the bottom of the ocean for some distance. Refined temperature devices are then used to record the difference in temperature between two points on the probe. By determining the thermal conductivity of the mud, it is possible to calculate the amount of heat flowing upward from the earth into the ocean water. Surprisingly, it turns out that approximately the same amount of heat is flowing from the interior of the earth in the oceanic areas as in the continental areas; namely, about 1.2 microcalories per square centimeter per second.

From all these figures it can be calculated that the average continental crust down to a depth of 35 kilometers produces about ½ microcalories per square centimeter per second from the radioactive breakdown of uranium, thorium, and potassium. This is about one half of the observed heat flow to the surface. It means that only about one half of the heat flowing to the surface comes from a depth of greater than 35 kilometers.

If you measure the amount of heat flow and estimate the thermal conductivity of the materials in the crust and below the crust, it is possible to estimate the increase in temperature with depth.

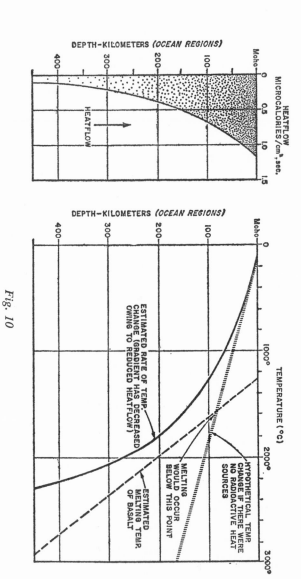

Fig. 10

66

Fig. 10 shows some estimates of the temperature-depth relationship. Note that the production of the heat in the crust greatly reduces the thermal gradient (rate of heat flow) at depth. It follows that the temperature at depth is very much less than would be expected if one simply measured the temperature in deep mines or other openings in the earth near the surface and extrapolated this information to great depth. In fact, if there were no radioactivity in the crustal rocks, the observed temperature gradients at the surface would require that the mantle be molten at a shallow depth; this is not in agreement with the known geological stability of the region. The facts, therefore, support the hypothesis that the radioactive components of the earth are largely concentrated in the near-surface layers.

By calculating the temperature at which materials would be molten at depths of 100 or 200 kilometers, it is possible to estimate how much radioactivity must be in the near-surface rocks in order to keep the temperature gradient within

Fig. 10. (opposite) *The temperature gradient, or maximum rate of change of temperature in a body, is proportional to the heat flow and inversely proportional to the conductivity. Near the surface of the earth, for example, the heat flow is 1.2×10^{-6} cal/cm^2/ sec; the conductivity is .007 cal/cm/degree C/sec. So the gradient is $1.2 \times 10^{-6}/.007 = 17 \times 10^{-5}$ degrees C per cm or 17° per km. Heat-producing elements lie between the surface and any point below; the heat flow at depth is therefore less and so is the gradient.*

known bounds. Attempts to do this have indicated that at least 0.2 part per million of uranium and 0.7 part per million of thorium on the average must be in the rocks down to 100 kilometers depth under oceanic regions. Since these amounts of radioactive elements would supply much of the observed heat flow to the surface, there must be little heat left flowing from the interior.

Thus we arrive at two important conclusions. The first is that very little of the original heat stored in the earth at the time of its formation is being lost, and, therefore, the earth is not cooling down at an appreciable rate, if at all. Secondly, we conclude that the major part of the earth's heat is coming from the breakdown of radioactive elements. Since almost all the breakdowns occur within the near-surface regions of the earth, it is reasonable to infer that some process has moved the radioactive elements to this location from a presumably homogeneous distribution at the time of the earth's origin.

Migrations through the Mantle

Again we see the need for some process to have brought up from within the earth the materials that make up the oceans and atmosphere and the radioactive elements. In support of this requirement, we see that uranium, thorium, and potassium do in fact belong to the group of elements that form compounds of rather low stability and therefore would be most likely to move to the outer part of

68

the mantle in any process in which partial melting was involved.

It is interesting that these conclusions do not violate the concept of an earth that is composed of materials similar to the iron and stone meteorites. The proportion of radioactive components in iron meteorites is very small indeed and would contribute a negligible amount to the heat of the earth if it were made of similar material. The amount of radioactive components in stony meteorites has been measured carefully, and it is a striking coincidence that the amount corresponds very closely with that needed to give rise to the observed heat flow in the earth if it were uniformly of stony meteoritic composition. The fact that the radioactive components have migrated to the outer part of the mantle does not alter this interesting and supporting evidence.

Thus we see an earth in which the central part is rather slowly changing, if at all, in temperature and losing heat to the outside very slowly. Near the surface, however, there is an important balance in which the heat produced by radioactive elements can flow to the surface without causing melting unless the system is disturbed in some way. If at the margin of a continent the accumulation of sediments formed a low-conductivity blanket which also contained added amounts of heat-producing elements, this might be enough to cause melting at a depth of 100 or 200 kilometers and give rise to volcanic activity and other effects related to mountain building.

There has been much discussion and difference

69

in opinion about the possibility of major convective overturns in the mantle down to the core boundary as a result of inhomogeneous distribution of heat sources. This process of convection could give rise to surface activity also and could be the cause of mountain-building events. In either case it is the heat from radioactivity that provides most of the energy for the dynamic events that have occurred at the earth's surface throughout geologic time.

CHAPTER IV

Measurement of Absolute Geologic Time

Since the first hot days when the thickening atmosphere swept in gales across the earth, alternately drenching it with rain and drying it with searing winds, its surface has been carved and molded, buried and exposed again. Unlike the moon, which has no atmosphere and has retained its primordial features until today, the earth has developed a challenging complexity of eroded surfaces and filled basins. The energy for these superficial alterations has come from the sun, and the agency has been the atmosphere and the water it contains.

Prior to the systematic study of geology, it was a continuing source of wonderment and superstition that creatures once living should now be found as parts of solid rock, and layers of sea shells found as ridges of high mountains. Today we take these evidences as proof of the dynamic movements of the earth's surface, of erosion and sedimentation, of the evolution of living forms. We piece together the information until we have a pattern that fits all the data. We can trace gradual change in the fossil remains of plants and animals as we study older and older beds, and notice how species converge to simpler common ancestors, until finally we get back to a time when there is no sign of life. We study the sequences of bedded rocks, giving them names and cataloguing them according to the extinct animals and plants that were living at the time of their deposition. When these sedimentary beds become compressed into rocks and are associated with a mountain-building event, we place this event, or *orogeny,* in its correct place in our story.

Relative Time Scale of the Planet's History

It was a major turning point in earth science when, in 1799, "Strata" Smith in England announced he had discovered that sedimentary strata of the same age consistently showed the same types of fossils. He demonstrated a complete sequence with proof that the succession of fossils was always in the same order. By 1808, French geologists had studied the fossils in the sedimentary beds of the Paris basin sufficiently to be able to correlate them with the strata worked out in England. The worldwide study of historical geology had begun.

These studies led eventually to the theory of evolution and great changes in man's approach to the historical aspects of the universe about him.

72

No longer was dogmatic creed to be superposed, by force if necessary, on facts denied by much material evidence. Centuries of superstitions fell before the challenging and rapidly advancing hypotheses about all manner of natural phenomena related to the planet earth and its relation to the universe.

In addition, these studies led to extensive interpretations of past advances and recessions of the seas on continental areas. This is natural, since the fossils were largely of creatures that lived in water and had durable parts, such as shells, and that were best preserved in sedimentary materials like muds and silts deposited in shallow seas. Since a considerable proportion of the continental areas is covered with sedimentary rocks, these studies have been of practical importance—particularly since accumulations of petroleum occur in such sedimentary cover rocks. However, had a means of correlation become available, it is probable that volcanic and mountain-building events would have been the more prominent features of the geologic time scale.

A fairly typical cross section of continental rocks is shown in Fig. 11, see also Plate I. This illustrates a gorge of the Colorado River. The crystalline rocks of the so-called *basement complex* represent the roots of an old mountain belt, which was eroded flat before the sea transgressed the land and deposited a layer of sediments. This eroded surface is known as an *unconformity*. The layer of sediments on it was then uplifted and beveled off by erosion, making a second unconformity before

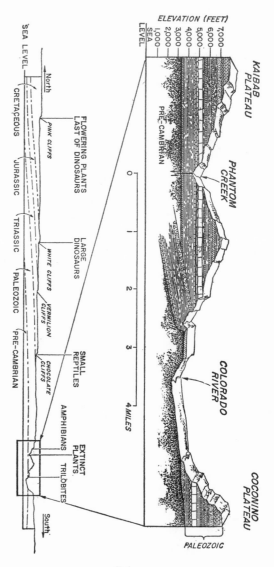

Fig. 11

74

the sea came in again and deposited the upper beds. Now crystalline rocks, the roots of the old mountain belt may originally have been formed of sedimentary materials, folded and crumpled by compression and buckling of the earth's crust. From the complexity of the structure it can be seen how difficult it is to select which transgression of the sea, or which geologic event, is the important one deserving a place on the area's geologic map. Actually, the material exposed at the surface is mapped, because information on sub-surface material is generally lacking. It is possible that the mountain-building event characterized by the crystalline basement rocks is the most important happening in this area, as it represents a time at which this part of the continent probably first emerged from the ocean depths.

Except in regions where there is no sedimentary cover, a geolgic map rarely shows any but the more recent events. The complexities make it too difficult to do otherwise.

In summary, then, we see earth history expressed at present largely in terms of transgressions of the sea and in the resulting thicknesses of sediments related to the fossil remains of living forms

Fig. 11. (opposite) The Grand Canyon is so deep because some great force from below lifted the entire region from sea level to more than a mile in altitude. Before this uplift the original land surface, shown as the Pre-Cambrian, was under water and covered by easily eroded sediments. The great depth permits the geologist to study strata and fossils that were marine in origin.

which were evolving through the geologically more recent past. The greater events in which mountain belts rose in globe-circling chains are left largely as gaps in the sedimentary record, because the method for dating and correlating them was not developed. With the advent of rapid and precise means of age measurement by radioactive methods, this is no longer so; the history of the earth in the future will more truly cover both the great expanse of time involved and the logically more important happenings.

Principles of Geologic Time Measurement

We have seen that the earth has had a history of large surface events such as the development of systems of mountains and areas of continents. These occurred at different times in different regions. Now only the roots of the mountains are left, owing to erosion in the ages that have elapsed. These mountain systems appear to have added continental areas to the earth's surface at the expanse of the ocean floor, forming stable blocks that do not sink, once they have come into existence.

The measurement of geologic time therefore includes the measurement of the age of the old mountain systems that make up the different parts of the continental crust. So, what we really are measuring is the time that has elapsed since the minerals in granites and other rocks crystallized. Our measurements will reflect one of two events: the time when molten materials came up into the crust, or the time when crustal materials were so deformed and heated that they recrystallized into their existing forms. Also we are able to determine the age of the earth itself (within fairly narrow limits) and the age of the elements that make up our part of the universe, by methods that differ from those used on crustal materials, as we shall see later.

In the later stages of earth history geologists are concerned with measuring the distinct periods in the evolution of living forms and the times when sedimentary sequences were deposited. This is more difficult, and in general we must seek the answers by bracketing the events with other measurements dependent upon the crystallization of some mineral containing a radioactive element.

In a crystalline rock such as a granite (see Plate II) the individual mineral grains are formed at a high temperature. Each crystal is not a pure chemical compound, although the components do come together in approximately the standard proportions for the particular mineral. All mineral crystals contain varying amounts of impurities: atoms of other elements that do not properly belong in the particular crystal structure. Because small amounts of almost all elements occur in all kinds of minerals, it is possible to measure the radioactivity of almost all mineral samples, whether the essential component of the material is a radioactive element or not.

For example, the ordinary mineral grains in a body of granite contain measurable amounts of uranium and thorium. These elements enter the

77

crystal structures during the latter's growth by occasional substitutions for the correct atoms, or by simple inclusion in imperfections of the growing crystal structure.

When a crystal is formed, it is a tight system that sometimes lasts for billions of years without disturbance. On other occasions the system is not tight and certain elements may diffuse through it or exchange with other elements in the environment. Thus, we must be careful in choosing mineral types with which to measure geologic age.

The methods of measuring geologic time by radioactivity require that a parent radioactive element be present in one of these tight crystal structures, and the crystal must have remained a closed system throughout the length of time to be measured. As the radioactive parent element undergoes spontaneous disintegration, it builds up a daughter product which provides a measure of the time since the crystal was mineralized. For example, if a pure crystal of uraninite, UO_2, contained no lead at the time of its formation, the breakdown of uranium into lead would start building up lead from zero at a rate depending upon the decay constant of the uranium. If this is known, and the content of uranium in the mineral sample, as well as the content of the lead daughter product, is measured carefully, the time required to build up the measured amount of lead can be calculated. If, on the other hand, the crystal structure contained some lead when the crystal was formed, the increasing amount of lead did not start from zero, and it is not possible to calculate the age unless

78

some other measurements are made. Fortunately, there is a way of doing this. I shall describe it later on.

Since it is not necessary that the mineral chosen for age measurement be one in which the parent element is a major component, it is possible to make age measurements on a mineral such as mica, which has a small content of rubidium as an impurity in the structure. One of the isotopes of rubidium is radioactive Rb^{87}, which breaks down to form an isotope of strontium Sr^{87}. Again, if there is any common strontium present in the crystal at the time it was formed, which generally is so, a special method must be used to determine how much was there.

Parent and Daughter Isotopes

The measurement of geologic time, therefore, depends upon finding a crystal structure of a mineral so resistant to change that it will remain essentially intact during its entire lifetime, and it must have a measurable quantity of some radioactive parent element, a measurable quantity of a daughter product forming from the disintegration of this parent, and in some way we must be able to determine the amount of daughter product present originally in the crystal structure. Furthermore, there must be no loss or gain of either the parent or daughter element throughout the life of the crystal, by any process such as diffusion or recrystallization or contamination from the surrounding region. It is truly amazing to think that in a rock

79

such as a granite there are little grains of minerals throughout the entire rock, each of which is a minute clock capable of telling the exact time since the formation of the rock.

Looking into the question in a little more detail, we see that we are dealing with radioactive isotopes rather than elements, and that the daughter products are also isotopes of some other element. For instance, U^{238} breaks down to form an isotope of lead Pb^{206}; U^{235} breaks down to form Pb^{207}; and Th^{232} forms Pb^{208}. Among the other important pairs are Rb^{87} breaking down to form Sr^{87}, and K^{40} (potassium) forming A^{40} (argon). Although normally the relative abundances of isotopes are constant, we find that among radiogenic isotopes (those isotopes that are formed from the breakdown of some radioactive parent) there is a considerable change in relative abundance.

For example, the isotopic abundance of common lead is approximately as follows:

Pb^{204}	1.4%
Pb^{206}	26 %
Pb^{207}	21 %
Pb^{208}	52 %

If, however, we measure the isotopic abundance of lead in an old crystal of uraninite, we find a much greater amount of Pb^{206} present; in fact, if there were no lead at all in the crystal structure to start with, all the Pb would be the 206 and 207 isotopes derived from the uranium isotopes present. Therefore, the relative abundance of radio-

Plate 1. A river can cut into rock only when there is a gravitational gradient to drive the water; thus, the steeper the terrain, the greater the power of the water to erode the rock. The Grand Canyon resulted from the uplift of a plateau originally at sea level to a great elevation in a relatively short time. The time since the uplift has been insufficient to develop the rounded forms of most topography. (Photograph by Paul Caponigro)

Plate 2. A slice of granite no thicker than paper between two pieces of Polaroid looks like this under the microscope. The rock is made of crystals of various minerals.

Plate 5. The story of evolution has grown from the study of the fossilized remains of living organisms preserved in rocks. This picture shows *Paradoxides bohemicus*, a trilobite. (Photograph by R. Paul Larkin)

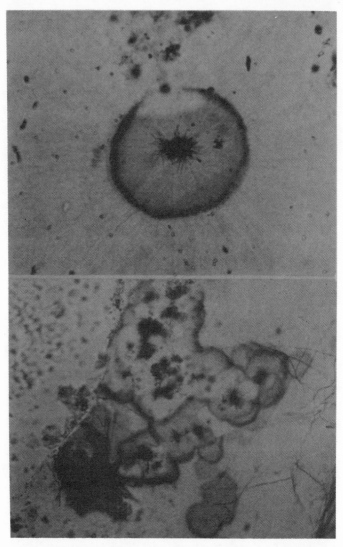

Plate 4. These are microscope photographs of the oldest known remains of living forms. They resemble some species of modern Algae. They have been studied by E. S. Barghoorn, of Harvard University, and S. A. Tyler, of the University of Wisconsin, and dated as 1600 million years old by the author. (Photograph courtesy of Professor Barghoorn)

Plate 3. The geochronology laboratory at the Massachusetts Institute of Technology. (Photograph by R. Paul Larkin)

Plate 6. The moon is believed to have formed at the
same time as the earth as a separate condensation
nucleus. Composed of the same material, it would
have heated by radioactivity enough to cause melting
in the interior, and therefore probably had volcanic
eruptions. The eruptions may have caused many of its
surface features. (Photograph by Harold M. Lambert)

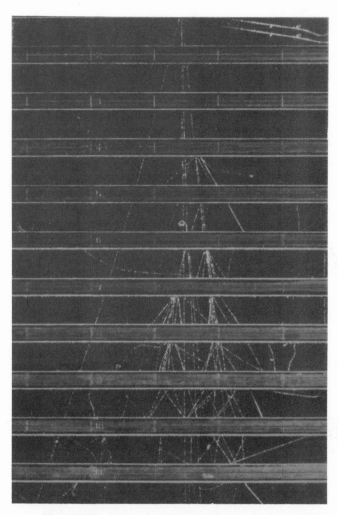

Plate 7. We cannot escape the radiations caused by the high velocity particles that rain on our atmosphere and are known as cosmic rays. As the tracks in the picture show, they can penetrate a considerable thickness of brass. Cosmic ray damage represents about a quarter of the total radiation damage we suffer continuously from all natural sources, exclusive of man-made sources. (Courtesy of Bruno Rossi and R. B. Brode of M. I. T.)

Plate 8. Half Dome is an example of giant processes at work. The battle between uplift and erosion stages a mighty scene at Yosemite. (Photograph by Paul Caponigro)

genic isotopes can vary considerably, depending whether or not these isotopes have been in close association with the parent element over a long period of time.

Thus it can be seen that there should be a normal slow change of the abundance of radiogenic isotopes with time in the earth's crust corresponding to the normal average slow breakdown of the total amount of parent in the crust. For example, surface rocks are estimated to contain about 0.035 per cent of rubidium and about 0.022 per cent strontium. The isotopic abundance in each element may be shown in the following way (Fig. 12). The radioactive rubidium 87 isotope breaks down at a rate such that half of it would have gone in 50 billion years, or about 5 per cent in the lifetime of the earth. Its breakdown product is strontium 87. Because rubidium 87 appears to be six times more abundant than strontium 87 in the earth's crust, the strontium 87, by the addition of the radiogenic material, should have increased by 30 per cent during the lifetime of the earth.

Actually, measurement of very old strontium minerals which do not contain much rubidium has shown this not to be so. The change is much smaller. The explanation is this: Rubidium is greatly enriched in the crustal rocks, and the actual amount in the upper mantle is very much less than the amount of strontium. This is interesting, because it provides us with another way of studying the relative enrichment of elements in the upper mantle and crust.

In a similar manner, we gain information about

the origin of the atmosphere by considering the enrichment of radiogenic argon 40 in it. The argon 40 in the present atmosphere is equal to about 25 per cent of the estimated total argon 40 produced by the breakdown of K^{40} in the earth since its beginning. The gases released during the development of the continents, which cover only about 20 per cent of the earth's surface, plus oceanic volcanic islands, would completely account for this argon

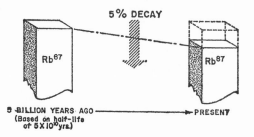

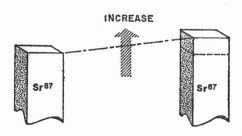

Fig. 12. Radioactive nuclides decay at known rates. Knowing how much of a given nuclide—rubidium 87, for instance—a mineral has now and its decay rate, we can calculate how much there was at any time in the past. Similarly, because rubidium 87 breaks down to form strontium 87, its change also can be calculated.

40. However, this rules out the possibility of much escape of gases from the mantle 4.5 billion years ago, or at the time the core and mantle separated.

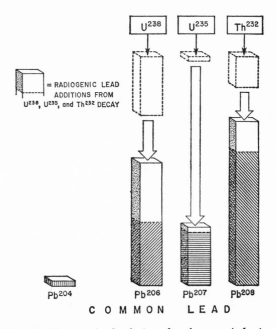

Fig. 13. *The standard relative abundances of the isotopes of common lead are shown above. As lead 204 is not changing with time, the standard ratios of common lead isotopes 206, 207, and 208 relative to lead 204 can be subtracted from any measured isotope distribution found in a uranium or thorium-rich mineral to give the true amount of radiogenic lead 206, 207, or 208 formed from the breakdown of the uranium or thorium. This will correct for common lead present in the mineral at the start.*

8 3

This is evidence in support of the hypothesis that the oceans and atmosphere increased slowly, keeping pace with the growth of the continental crustal material.

Looking now at the problem of determining whether there was any of the daughter element present in a mineral at the time of its crystallization, we see that we can make certain assumptions regarding the probable isotopic composition of the daughter element at the time the mineral was formed. If you will refer back to our example of common lead present in a crystal of uraninite, Fig. 13 indicates how it is possible to correct for this element by the measurement of the isotopic abundance of the final lead now present in the sample.

For this calculation we use the common lead ratio of the lead abundance table. The amount of non-radiogenic Pb 204 indicates the amounts of Pb-206, 207, and 208 in the mineral at the time of its crystallization. When we subtract these amounts from the measured abundances, we have left the amounts of radiogenic Pb-206 and 207.

Contamination of samples in the course of analysis introduces another difficulty into the process of determining geologic age through radioactivity measurements, but we use refined methods of correcting for the contamination. A little further on I shall go into this in some detail, and with some simple mathematics, because a concrete example will illuminate the technique of age measurement and at the same time show the relationship between certain forces involved in these complex experiments and certain forces studied in simple experiments performed in every high school physics laboratory.

The accuracy attained in these contamination corrections is truly amazing. Let us suppose, for example, that we have a young mineral containing argon 40 developed in the breakdown of potassium 40. Although each gram of the mineral sample contains only 10^{-5} cubic centimeters of argon 40 under standard conditions of temperature and atmospheric pressure, it is possible to separate the argon from the sample, to purify it and measure it. But the amount is so small that any air contamination remaining in the sample or any air leak in the course of analysis would destroy the accuracy because all samples of air contain a small amount of argon 40. A correction is necessary. Fortunately, the air also contains the argon isotope A^{36} in such proportion that the ratio of A^{36} to A^{40} equals .00337. So, if we measure the amount of A^{36}, we can calculate the amount of A^{40} that is air contamination. When we subtract that amount from the total argon 40 in the mineral sample, we can determine the true amount of radiogenic A^{40} in the sample, despite the slight A^{40} contamination from the air. In very young materials it is almost impossible to eliminate all contamination; to obtain accurate age measurements, the correction is essential.

History of Age Measurements

But before we go into the techniques in detail, let us review briefly the history of this fascinating branch of science. It is important because the development of accurate geologic time measurement coincides with the growth of our knowledge of radioactivity.

It all started, quite by chance, when the French physicist Henri Becquerel (1852–1908) left a photographic plate and a piece of a uranium salt together in a drawer. The next time he looked, something had darkened the plate; in other words, it had been exposed to some unknown radiation. This fortuitous incident turned Becquerel and his contemporary experimenters onto a new track that led ultimately to grand new achievements in physical science: our knowledge of the atomic nucleus, radioactive disintegration, nuclear fission, and fusion.

Soon after Becquerel's discovery, Pierre and Marie Curie began their long and now famous search for sources of radioactivity and eventually succeeded in isolating radium and several other radioactive elements. The Radium Institute in Vienna and other organizations joined in the search for radioactivity in other elements, and it has continued to this day.

It now is known that all atoms with Z-number above 82—that is, whose nuclei contain 82 or more protons—are radioactive, and many naturally occurring lighter ones are also. The heavy elements were found to be related to three natural radioactive series. These are the uranium series, the thorium series, and the actinium series. The uranium series is shown in Fig. 9. The other two series are of a similar nature. In all, there are about sixty known naturally occurring radioactive nuclides. Most are short-lived and exist only because the breakdown of some radioactive parent element or some other nuclear reaction forms them continuously.

The application of research in radioactivity to understanding of the earth and universe had its beginning shortly after the turn of the century. Much had been learned about the decay scheme of uranium, the constancy of the ratio of uranium to radium, and it had been demonstrated that lead was the stable end product of the disintegration series. It soon became apparent that the radioactive process involved the generation of heat and thus was of great significance in the study of earth history.

In his book *Radioactivity in Geology* (1909), John Joly opened the way to new understanding of earth heat and the effect of radioactivity on the time required for the earth to cool from its formation temperature. The nineteenth-century authority on thermodynamics, Lord Kelvin of Glasgow University (1824–1907), had studied the cooling question and come to the conclusion that the earth was only 24 million years old. But Joly demonstrated that radioactivity in the rocks changed the picture altogether. (He also was the first to recognize the possibility that radiogenic heat—heat generated in radioactive disintegrations —supplied the energy for the great mountain-building processes.)

Another discovery was to be of much interest in the question of earth age. This was that helium, first found on the sun and then on earth, was a product of the breakdown of uranium and thorium. It was another of the tools that, early in this century, enabled scientists to make their initial attempts to measure earth age through the process of radioactivity. The great Lord Rutherford was the first to study the relation of helium to uranium in minerals, but the helium loss in the highly radioactive minerals he measured was so large his results indicated very low ages.

After Lord Rutherford's work with helium came systematic investigation of lead-uranium ratios, and this research quickly created a completely new dimension in geology—the approximate measurement of geologic time in absolute terms. The period of intensive study of lead-uranium and lead-thorium ratios in radioactive minerals of various ages had reached a climax in 1931 with the publication of *The Age of the Earth,* by Knopf, Schuchert, Kovarik, Holmes, and Brown, in the National Research Council's "Physics of the Earth" series.

In brief, this line of investigation made two significant findings: that a fairly orderly progression of ages back to about 2 billion years in various parts of the earth could be demonstrated, and that this progression was in agreement with the geological interpretation of the succession of events. The radioactivity studies corroborated the older interpretations which showed that certain mountain belts were older than other belts cutting through them. But the data were sketchy, yielding only a few points on which to hang all geological history, and there were many *discordances,* or failures of independent lines of investigation to produce results in agreement with each other. What caused most of the trouble was the presence of common lead that had existed in the minerals at the time they were formed. A new approach was needed.

More powerful new methods were found, in the period 1935 to 1940, in the study of isotopes and development of mass spectometry, which is a technique for determining the masses of positively charged particles and for finding the relative amounts of the isotopes in an element. In classic papers published from 1938 to 1941, A. O. Nier, then at Harvard University, opened up the possibility of making precise age determinations by separate isotope measurements and by making corrections for the presence of daughter elements at the time of crystallization of a mineral. Gains or losses of the elements of interest could be detected by means of *concordancy* of different isotopic ratios in the same sample.

This period also saw the first publications on artificially produced radioactivity, and the uses of this tool in geological science advanced rapidly. In 1935, Nier demonstrated the existence of the radioactive isotope K^{40}. About the same time, in Berlin, a team of four scientists—Lise Meitner, Otto Hahn, Fritz Strassman and Otto Frisch—was carrying out a series of perplexing experiments that led eventually to the discovery of nuclear fission. The age of experimental nuclear physics had begun.

89

It was in the same period that Charles S. Piggott and William D. Urry were studying the deposition of ionium, a daughter product in the breakdown of U^{238}, in ocean-floor sediments. It was found that sea water deposited ionium in excess of its equilibrium amount of parent element uranium; as subsequent deposition of sediments buried the ionium, it decayed with an 84,000-year half-life. Measurements taken in the topmost few centimeters of ocean-floor sediment showed that the radioactivity gradually diminished with depth. Under ideal conditions, this could be interpreted to give the rate of deposition of sediments, a measurement of considerable value to the understanding of the geology of these vast areas, and to the investigation of the abundance of elements in the earth's crust.

After studying under Nier, at the University of Minnesota, L. T. Aldrich, of the Carnegie Institution, became interested in geochronology, and, together with G. W. Whetherill, at the Department of Terrestrial Magnetism of the Carnegie Institution, led the important investigations of the ratios of argon 40/potassium 40 and strontium 87/rubidium 87 in minerals. Since Nier's original work great strides have been made in the study of radiogenic isotopes of lead and their variations, both in measuring the age of the earth and meteorites and in seeking better understanding of the relationship between the continental crustal rocks and the mantle below them. And of major importance in archaeology and the geology of "recent" events— a few tens of thousands of years ago at the most—

90

has been the development of the carbon 14 method of analysis by W. F. Libby and his associates at the University of Chicago.

Many other avenues of study in the fields of age measurement and radioactivity have been opened up, but it is beyond the scope of this book to consider them all in detail. In this brief survey of the progress of the investigations we have tried to present a broad picture of the problems involved and to convey a general idea of what kinds of events are being measured. Now we are ready to get down to specifics.

Techniques and Instruments

First, we must examine the physical principles underlying the measurements needed in age determination. As we have seen, it is necessary to measure the absolute amount of a single isotope in a sample of mineral or rock. Suppose, for example, that we are measuring the A^{40}/K^{40} ratio in a sample of black mica (biotite) from a granitic region in the Northern Territories of Australia.

The mica minerals are particularly good for this method of age determination because they contain much potassium and are easily concentrated from the rock. Geologists collect a sample of the granite in the field and take it to the laboratory, where it is crushed and ground and the mica separated from it. A sample of the mica is then analyzed chemically (or by flame photometer or other standard procedure) for potassium. For refined work this measurement is made by mass spectrom-

91

eter. Analysis gives the total amount of potassium in the sample, and since the isotope K^{40} bears a constant relationship with the other isotopes of potassium, the amount of K^{40} can be calculated. Actually, the proportion of K^{40} in total K is .000119 per cent.

Another sample of the mica is then weighed and placed in a furnace which is part of a vacuum-tight system such as shown in Plate III. Almost all traces of air contamination are removed from the system and from the sample by warming the sample and pumping out all of the gases for a period of time. The furnace is then heated to such a temperature that the sample of mica melts and the argon which has formed from the decay of K^{40} is released into the system. If no other argon from any air contamination were present, this argon would all be radiogenic A^{40}. But if there are traces of air contamination still remaining either in the sample or in the system, there will be a small amount of A^{36} which always accompanies air argon in a ratio of about one atom in 300.

In order to measure the absolute amount of A^{40} released from the sample, it is necessary to add a known amount of another isotope of argon against which the abundance of A^{40} can be compared. Generally A^{38} is added. (An isotope used for this purpose is known as a "spike.") A known number of atoms of A^{38} is metered into the apparatus through a system in which the volume can be measured very accurately. Most of the gases from the sample are then removed from the system with a material such as hot titanium metal, which absorbs most gases readily. Hydrogen is removed by oxidizing it in the presence of heated copper oxide and then condensing the water that is formed at liquid nitrogen temperature, along with any other gases that may condense at the same temperature. Almost pure argon is left in the system, and this is now composed of three different quantities. First there is the radiogenic A^{40}; next there possibly is a small amount of argon from air contamination (A^{40}, A^{38} and A^{36}); and thirdly there is the known amount of A^{38} spike. This mixture of argon isotopes is now ready for measurement in a mass spectrometer, which determines the ratio of the isotopes 40, 38, and 36.

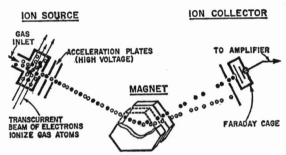

Fig. 14. An ion of mass m_1 and electric charge e will be separated from another ion of mass m_2 and the same charge e as it passes through a magnetic field, if traveling the same initial path. This is the principle of the mass spectrometer.

A diagram of a mass spectrometer is shown in Fig. 14. The argon is admitted to the instrument through a small opening; commonly only about 10^{13} atoms pass through this opening per second by diffusion into the evacuated part of the spectrometer. As the individual atoms of argon enter into the mass spectrometer, they meet a transcurrent beam of electrons generated by a hot filament and pulled across their path by an electric field. Most of the atoms of argon are ionized by the electrons; this means that when an electron strikes a neutral atom of argon it removes one of the atom's outer electrons, leaving it with a single net positive charge. The argon atom, then known as a positive ion, can be moved in space by placing it in an electric field. This electric field is present in the volume where the gas is being ionized; as soon as an ion is formed, it is accelerated by this electric field and obtains a velocity dependent upon the strength of the electric field.

For example, if an ion is accelerated through an electric field with a total potential of 2000 volts, the ion will attain a velocity that can be calculated as follows:

If a particle carrying an electric charge equivalent to one electron (that is, a single ionized particle) is accelerated through a voltage V, by definition it will have a kinetic energy of V electron volts. The electron volt has been determined through experiment to be equivalent to 1.60×10^{-12} erg. In this example, therefore, the kinetic energy imparted to the argon 40 ion is $2,000 \times 1.60 \times 10^{-12}$ erg.

Since we know that kinetic energy equals one half the mass (m) times the square of the velocity (v), or

$$K.E. = \tfrac{1}{2}mv^2$$

we can proceed to solve the equation for v after we have calculated the mass of the argon 40 ion. To do this, we must go back to one of the milestones in the progress of physical science, to an hypothesis advanced by Amadeo Avogadro (1776–1856). One of Avogadro's conclusions was that there are 6.02×10^{23} atoms in an amount of any element equal to its atomic weight in grams. The atomic weight of A^{40} is 40; so 40 grams of A^{40} will contain 6.02×10^{23} atoms, or one atom of A^{40} will weigh $\dfrac{40}{6.02 \times 10^{23}}$ grams. (The constant 6.02×10^{23} is known as Avogadro's Number and applies also to the number of molecules in a weight of a substance equal to its molecular weight.)

So, substituting in our K.E. equation, we have

$$2000 \times 1.60 \times 10^{-12} = \tfrac{1}{2} \times \frac{40}{6.02} \times 10^{-23} \times v^2$$

or

$$v = 0.98 \times 10^7 \text{ cm/sec}$$

or

$$= 216{,}000 \text{ miles per hour}$$

The velocity of an argon 38 ion, being of lower mass, is greater. Because

$$v^2 = \frac{2(\text{K.E.})}{m}$$

The ratio $v_1{}^2/v_2{}^2$ for two different ions is inversely proportioned to their masses

or

$$v_1{}^2/v_2{}^2 = m_2/m_1$$

9 5

so that

$$v_1/v_2 = \sqrt{m_2/m_1} \text{ or } v_2 = v_1\sqrt{m_1/m_2}$$

Therefore the argon 38 ion would be traveling

$$0.98 \times 10^7 \times \sqrt{40/38} \text{ cm/sec}$$

or

$$1.00 \times 10^7 \text{ cm/sec}$$

A particle of mass m, with an electric charge e, moving at right angles to a magnetic field of strength H, will have a force acting on it equal to Hev, and it will cause the particle to travel in a circular path. Now we know from the mechanics of a weight on a string that for a circular motion in a path of radius r, there must be a force equal to mv^2/r acting on the mass toward the center of the circle.

Thus in the magnetic field the particle will take a circular path of such a radius that the magnetic force Hev equals mv^2/r, or the radius $r = \dfrac{mv}{He}$.

Again referring to our two ions, argon 40 and argon 38, which we will designate by subscripts 1 and 2, we see that

$$r_1/r_2 = m_1v_1/m_2v_2$$

But

$$v_1/v_2 = \sqrt{m_2/m_1}$$

so that

$$r_1/r_2 = \sqrt{m_1/m_2}$$

In other words, if the two ions are accelerated through the same voltage and pass through the same magnetic field, they will be deflected at different angles, because they will travel a short distance along circular paths of different radii as shown in Fig. 14.

Large numbers of these ions traveling along these paths are called ion currents, or ion beams. It can be seen that if the ion beam is measured when it enters a fixed slit beyond the magnetic field, a variation of either the accelerating voltage or the magnetic field will cause one or the other of the types of argon ion to be measured. This is actually done in practice, and the ion beams for argon 38 and argon 40 are alternately measured as they are moved back and forth across the pick-up slit by changing the magnetic field slowly. Thus the ratio of argon 38 atoms to argon 40 atoms entering the mass spectrometer can be determined accurately.

An example of a record showing the amplified current from ion beams of A^{38} and A^{40} taken during an age measurement is given in Fig. 15. The peaks get progressively smaller because the sample is being used up. In a normal analysis A^{36} is also measured to indicate possible air leaks or contamination.

Returning now to our example of an age measurement of mica from Australia, we see how it is possible to determine its age using the isotope dilution technique. Using an actual measurement as an example, the gas from a 1.81-gram sample was "spiked" with 10.95×10^{15} atoms of argon 38 introduced into the furnace from a measured volume.

The total gas was purified and the ratio of argon isotopes measured on a mass spectrometer. The measured ratio of argon 36 to argon 40 was found to be negligibly small so that it could be safely assumed that there had been no air contamination.

The measured ratio of argon 38 to radiogenic argon 40 was found to be 0.446, so that the actual

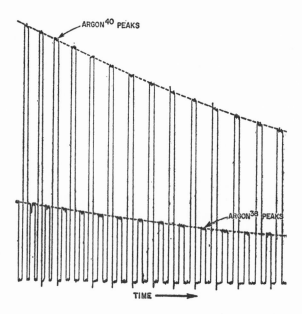

ARGON⁴⁰ PEAKS

ARGON³⁸ PEAKS

TIME ⟶

Fig. 15. The relative amounts of argon 40 and argon 38 are shown in this mass spectrometer record. Argon 40 is indicated by the taller peaks and argon 38 by the lower peaks. Both amounts decrease in time as the original sample is used up, but the proportional heights remain the same.

98

number of argon 40 atoms in each gram of sample was given by

$$\text{Argon 40 atoms/gm} = \frac{10.95 \times 10^{15}}{.446 \times 1.81} = 13.58 \times 10^{15}$$

The total potassium was found to be 4.21 per cent by separate analysis. But of this total potassium only .0119 per cent of the atoms are radioactive potassium 40, or in one gram of sample there were

$$.0421 \times .000119 \times \frac{\text{Avogadro's Number}}{\text{atomic weight}}$$

or

$$77.1 \times 10^{15} \text{ atoms of potassium 40}$$

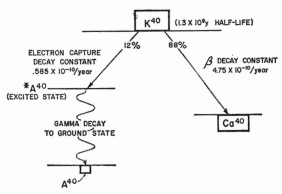

Fig. 16. Potassium 40 breaks down to calcium 40 by beta decay in 88 per cent of the cases, and to argon 40 by electron capture in 12 per cent of the cases, with the disintegration constants shown. This ratio is a statistical constant and does not change with time. The diagram also shows that argon 40 forms first in an excited state (A^{40}), which then transforms to argon 40 in the ground state by emission of a gamma ray.*

99

Therefore, 13.58×10^{15} atoms of argon 40 had developed from the radioactive breakdown of some number of potassium 40 atoms, of which there are now 77.1×10^{15} atoms left.

Potassium 40 breaks down to calcium 40 also (see Fig. 16) in a ratio of about 8.47 calcium atoms to one argon atom. This means that at the time of crystallization of the mica, each gram of it contained potassium 40 atoms equal to

$$(77.1 \times 10^{15}) + (13.58 \times 10^{15}) + (8.47 \times 13.58 \times 10^{15})$$

or

$$205.6 \times 10^{15} \text{ atoms.}$$

The rate of breakdown of a radioactive substance remains constant, but as the parent atoms decrease in number, the actual rate at which daughter atoms are produced also diminishes. This is similar to our example of the absorption of gamma rays earlier, and again we see an exponential relationship.

Let us look at this process in Fig. 17. Starting at any time, indicated as zero, when the mineral is formed, the potassium 40 starts to break down. It breaks down exponentially with time so that the amount left in the mineral sample is indicated by the curve marked K^{40}. In order to test this we can see from the curve that one half of the K^{40} is left after 1.31×10^9 years, which is the half-life of K^{40}.

In the same sample, however, the products of the breakdown, Ca^{40} and A^{40}, start building up from zero, and the total number of atoms of both combined equal the number of atoms of K^{40} that

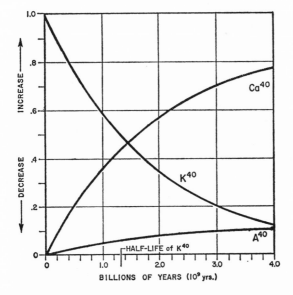

Fig. 17. A unit amount of potassium 40 breaks down exponentially with time with a half-life of 1.31 × 10⁹ years, as seen on the curve marked K⁴⁰. If this occurs in a closed system, the decay products Ca⁴⁰ and A⁴⁰ will start to build up in a ratio shown by the two other curves. The sum of these two nuclides at any time equals the amount of K⁴⁰ that has disintegrated.

have gone. Therefore, the sum of the curves marked K^{40} and $Ca^{40} + A^{40}$ always equals the original number of atoms of K^{40} existing at time zero.

Now we know that A^{40} and Ca^{40} always form in the same ratio. In other words, the probability of a K^{40} atom's decaying to an atom of Ca^{40} in a unit of time is a constant, and similarly the proba-

101

bility of its decaying to an atom of A^{40} is another constant, so that the ratio of the two products always remains the same. Thus we can draw curves marked Ca^{40} and A^{40} so that their ratio is a constant, and they both add up to the curve marked Ca^{40} and A^{40} at any time.

You can now see that the ratio A^{40}/K^{40} changes with time. If we measure this ratio in the mineral sample, we can find from the curve how many years have elapsed since the mineral was formed, at time zero. But in practice the age is calculated accurately and not taken from the curve, and we will go through this calculation to see how it is done. If M is the number of parent atoms left at any time t, and M_0 the number to start with

$$M = M_0 e^{-\lambda t}$$

where λ is the proportion of parent atoms breaking down in a unit of time.

For potassium 40, the decay constant λ is 5.28×10^{-10} per year. In our example $M_0 = 205.6 \times 10^{15}$ atoms and $M = 77.1 \times 10^{15}$ atoms, so that we can solve for t, the time required for this much decay of the potassium. This can be done by taking the common logarithm of the equation

$$\log M = \log M_0 - \lambda t \log e$$

or

$$16 \log 7.71 = 17 \log 2.05 - 5.28 \times 10^{-10} t \times \log e$$

As $\log e = .434$ the solution gives

$$t = 1860 \times 10^6 \text{ years, or the mica is 1860}$$
million years old.

In the measurement of Sr^{87} and Rb^{87} there is no satisfactory way to have these isotopes in the form of a gas, like argon. Samples of compounds of strontium and rubidium must be heated on a filament in the mass spectrometer. Ions of strontium or rubidium are emitted by thermal ionization, rather than created by electron bombardment. These ions are accelerated in the way we have described, and the measurement is otherwise the same. This type of instrument is known as a solid-source mass spectrometer. There are always small corrections to be made in these kinds of measurements; for example, the argon isotopes, because of their different masses, do not flow at the same rate through the small opening, and the strontium atoms come off the filament at a different rate according to their different masses. Generally we can calculate, or correct for, these "fractionation" effects.

Present-Day Methods

Different investigators have their own favorite methods of measuring geologic age, and all the methods have a useful place. At the present time it appears that two methods are emerging as most generally useful for most of the earth's history. These are the Sr^{87}/Rb^{87} and A^{40}/K^{40} ratios. In general these methods can be used on potassium-rich minerals, which are very common. Although the Sr/Rb method does not depend upon potassium, rubidium commonly is in minerals that are rich in potassium, and it is possible to compare results on the same mineral by the two different methods. Many common rock minerals, particularly the group of micas such as muscovite, biotite, and lepidolite are used. The potassium feldspar, orthoclase, may be used in certain circumstances.

Orthoclase does not yield satisfactory ages in the A/K measurement method, the results generally being about 25 per cent low. The cause of this discrepancy is not yet definitely known. The mineral seems to be usable for Sr/Rb ratios, however. This means that most crystalline rocks such as granites, gneisses, schists, and slates can be used in age determination because they contain satisfactory minerals. Of course, if rocks have been recrystallized, the age that is measured is only the age of the recrystallization.

Applying the A/K method and the Sr/Rb method to the same mineral gives a cross-check useful in determining whether the material is suitable for age measurement. If the two ratios agree, it is fairly good evidence that the daughter products have not been lost since the time of the crystallization of the mineral, because it is to be expected that such dissimilar atoms as A and Sr would be lost at different rates, if any leakage occurred.

One of the most difficult problems in the development of these methods has been the accurate determination of the half-life of the parent element. The values for the decay constants are still not settled, although temporary values are being used. However, the agreement in the ages determined for the same minerals by the different methods indicate that the half-life values must be nearly correct.

Still of major usefulness is the measurement of lead-uranium ratios in minerals that have a high content of uranium. Today in this method it is necessary that the three different age ratios Pb^{206}/U^{238}, Pb^{207}/U^{235}, and Pb^{207}/Pb^{206}, give ages that are in agreement before the age measurement as a whole can be considered acceptable. When this is the case, the age is said to be "concordant." Frequently a separate ratio can be obtained for the same mineral if there is a small amount of thorium present: namely, the ratio Pb^{208}/Th^{232}. All these separate measurements give information regarding the history of the mineral and its general acceptability. When a mineral has yielded a concordant age, this age is generally considered to be highly foolproof. These concordant ages have therefore been used to a large extent as key age points against which other methods can be checked. The drawback with the method is the lack of general distribution of radioactive minerals and the fact that these radioactive minerals have suffered such a large amount of radiation damage that they frequently show ages that are not concordant.

The Carbon 14 Method

Carbon 14 emits a beta particle with an energy of .15 Mev and has a half-life of about 5600 years. It can be created in the neutron pile from nitrogen 14 with thermal neutrons. In the atmos-

105

phere the neutrons are liberated by cosmic radiation and are most abundant above 30,000 feet. Most of the neutrons liberated by cosmic radiation eventually go into the formation of carbon 14 out of the nitrogen of the atmosphere, so that carbon 14 production, and also the rate of C^{14} decay, is almost equal to the rate of production of neutrons. The carbon 14 rapidly oxidizes to CO_2 and enters the carbon cycle in the surface of the earth. It is estimated that the total carbon in the atmosphere, oceans, and in the biological material on the earth's surface amounts to approximately 8.3 grams per square centimeter of surface.

If in the atmosphere 2.4 neutrons on the average are produced per square centimeter of earth's surface per second, the disintegration rate of the total carbon 14 should be $2.4 \times 60/8.3 = 17.2$ disintegrations per gram of carbon per minute. Actual measurements of the radioactivity of ordinary living carbon on the earth's surface today yield a mean value of 16.1 disintegrations per minute, which is in excellent agreement with the theoretical prediction.

When carbon in some compound becomes buried or removed from the living cycle of carbon, so that no further additions of carbon 14 enter into the compound, the carbon 14 that the compound started with begins to decay on a half-life of 5600 years. For example, if a piece of wood in the tomb of one of the Pharaohs of Egypt was obtained from a tree living at the time, and since that time had remained buried below the surface and out of the range of neutrons produced by cosmic rays, it

106

would have measurably less carbon 14 than a piece of wood living today.

As we have said, this method developed by Libby and his co-workers has proven to be highly successful in archaeological research. In order to demonstrate that the method worked, as well as

AGE DETERMINATIONS ON SAMPLES OF KNOWN AGE

SAMPLE	SPECIFIC ACTIVITY (cpm/g of carbon)		AGE (years)	
	FOUND	EXPECTED	FOUND	EXPECTED
TREE RING	10.99±0.15	10.65	1100±150	1372±50
PTOLEMY	9.50±0.45	9.67	2300±450	2149±150
TAYINAT	9.18±0.18	9.10	2600±150	2624±50
REDWOOD	8.68±0.17	8.78	3005±165	2928±52
SESOSTRIS	7.97±0.30	7.90	3700±400	3792±50
ZOSER: SNEFERU		7.15	4750±250	
ZOSER	7.88±0.74			4650±75
	7.36±0.53			
SNEFERU	7.04±0.20			4600±75

This classic set of measurements of the carbon 14 content of ancient pieces of wood of known age launched an important new method of dating the geologically recent past. Willard F. Libby, of the University of Chicago, developed the method; his associate, James R. Arnold, made these measurements. The objects included a tree trunk from an Arizona excavation, part of a coffin of the Ptolemaic period of Egypt, wood from a floor at Tayinat, Syria, deck board from the funeral boat of the Egyptian Sesostris III, and pieces of wood from the tombs of Sneferu of Meydum and Zoser of Sakkara, Egypt.

107

to calibrate it, the investigators obtained samples of carbon-bearing material from locations in which the age of burial was known fairly accurately. In this way the method could be tested. The preceding table shows the original set of determinations.

Since this work large numbers of analyses have been made of carbonate-bearing materials of all kinds—including shells from the ocean, carbon in deep ocean waters to determine the rate of circulation, peat in peat bogs, ancient relics of all types, and many things which have both geological and archaeological interest.

The actual measurement of the sample is not too difficult, except for the problem of reducing the background of radioactivity to a low enough level so that the weak beta activity of carbon 14 may be measured accurately. Solid or gas counting may be used; the sample may be a solid coating of carbon black inside a Geiger counter, or the carbon may be in a gas, such as carbon dioxide, when it is introduced into the counter. The counting chamber is surrounded by an iron or mercury shield, to remove the soft component of background gamma radiation, and a ring of anti-coincidence counters which correct for stray cosmic ray activity. These techniques usually reduce the background of the counter to about three counts per minute or so. Without rather special developmental work, it is not generally practicable to measure ages in excess of about 20,000 years, because the radioactivity of the carbon becomes so slight that it is difficult to get an accurate measurement above background radioactivity.

108

The carbon 14 method does not depend upon the presence of a long-lived radioactive isotope that has remained in existence since the origin of the elements. It depends upon the continuous formation of a short-life radioactive element, by what is believed to be a fairly constant nuclear process. Obviously, the method depends upon constant neutron flux into the upper part of the atmosphere or a fairly predictable quantity of carbon in the reservoir that makes up the cycle. Actually, a correction must be made for the increase in carbon in this carbon cycle during the last fifty years because of the large amounts of coal and oil that have been burned since the Industrial Revolution.

CHAPTER V

Memorable Dates in Earth History

Being egocentric creatures, we human beings are understandably inclined to start our examination of the universe by looking at ourselves, at the ground we walk on, at the mountains that tower above us, at the oceans that lap night and day upon our shores and spread their waters farther than the eye can see. Our "universe," therefore, tends to be a local universe, which we soon find, did not start all at the same time. We now must study each part of our environment separately before we can speak of the "age of the universe" with understanding. Later, we get back to the origins of the earth, our galaxy, and the elements of which these are made and, ultimately, to the universe of galaxies.

In this and the final chapter we shall try to set forth the best present answers to these tremendous matters. By stages we shall go from the time the earth's crust became stable—from the first fossil records of the earliest forms of life to the evolution of Homo sapiens—Man. Then, having established a platform and Man to stand on it, we shall let him look upward to ponder the greater problems of genesis. An increasingly well established advance base in this pioneer field of science is our knowledge of the age and origin of the earth.

So far measurement of the oldest rocks of the continental masses over most of the earth's surface has produced an unusually consistent set of numbers. The oldest age measured is approximately 2600 million years, and this value has been found in North and South America, Asia, Africa, and Australia. No older reliable age has been recorded in any other region.

Thus, geologists assume that the continental earth's crust was either too unstable before this time to have been preserved or else had not yet started to segregate from the mantle.

These earlier parts of the continents were probably surrounded by oceans not quite as deep as today's. We believe that these continental areas were not eroded down to a very much lower level than the present sea level, because, on the average, their surfaces are not submerged platforms but are unexpectedly close to the present sea level. One unexplained feature of interest is that most of the largest gold mines in the world occur in these areas.

Following these first mountains of 2400–2800 million years ago came a succession of mountain-building events lasting to the present. The distribution of these is indicated approximately in Fig. 18. The greatest mountain systems of all time have been the youngest, which have swept in great

111

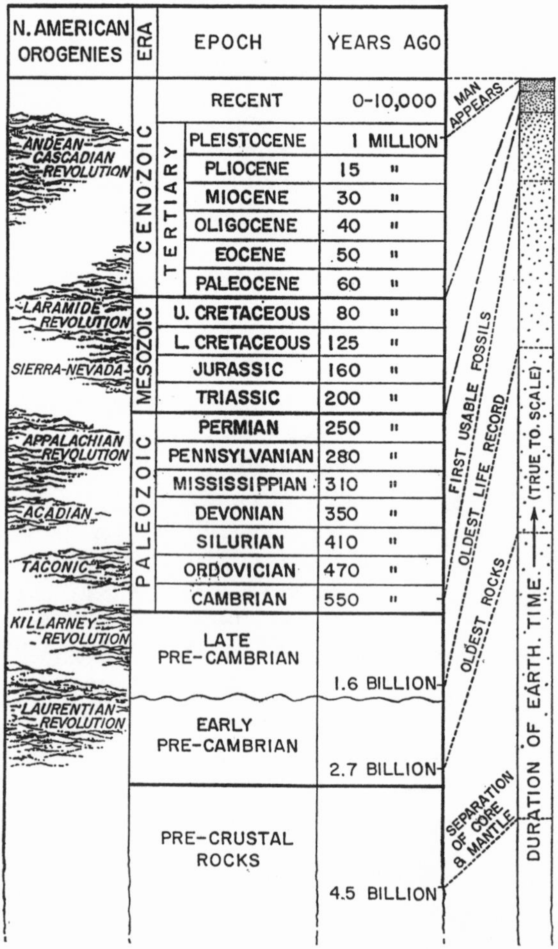

N. AMERICAN OROGENIES	ERA	EPOCH	YEARS AGO	
		RECENT	0-10,000	MAN APPEARS
ANDEAN CASCADIAN REVOLUTION	CENOZOIC / TERTIARY	PLEISTOCENE	1 MILLION	
		PLIOCENE	15 "	
		MIOCENE	30 "	
		OLIGOCENE	40 "	
		EOCENE	50 "	
		PALEOCENE	60 "	
LARAMIDE REVOLUTION	MESOZOIC	U. CRETACEOUS	80 "	FIRST USABLE FOSSILS
		L. CRETACEOUS	125 "	
SIERRA-NEVADA		JURASSIC	160 "	
		TRIASSIC	200 "	
APPALACHIAN REVOLUTION	PALEOZOIC	PERMIAN	250 "	OLDEST LIFE RECORD
		PENNSYLVANIAN	280 "	
		MISSISSIPPIAN	310 "	
ACADIAN		DEVONIAN	350 "	
		SILURIAN	410 "	
TACONIC		ORDOVICIAN	470 "	
		CAMBRIAN	550 "	OLDEST ROCKS
KILLARNEY REVOLUTION		LATE PRE-CAMBRIAN		
			1.6 BILLION	
LAURENTIAN REVOLUTION		EARLY PRE-CAMBRIAN		
			2.7 BILLION	
		PRE-CRUSTAL ROCKS		SEPARATION OF CORE & MANTLE
			4.5 BILLION	

(TRUE TO SCALE)

DURATION OF EARTH TIME.

Fig. 18

circles much of the way around the earth. One belt of these follows the western edges of North and South America and continues on the western side of the Pacific. Another belt goes almost at right angles from the East Indies through the Himalayas and the Alps.

It is interesting to note in these historical maps that there is some tendency for the continents to have grown outward from central nuclei by the addition of belts of mountains at the margins. This is not always true. Some of the oldest mountains terminate sharply against an ocean. Not enough work has been completed to answer all the questions yet, and there are many interesting ideas to be explored. For example, one of the boldest ideas put forth has been supported by a group of European geologists led originally by Alfred Wegener. This hypothesis has the continents drifting apart from a central single land mass.

Life and Evolution

What about life on earth? How long did it take for evolution as we think of it to produce a man from the simplest forms of life?

Remains (imprints of living forms) described by Elso S. Barghoorn and his co-workers at Har-

Fig. 18. (opposite) The continents are made up of ancient, eroded mountain belts, like logs in a raft. These developed at different times in earth history, from a measured 2600 million years ago to the present. The geological periods are related to these great mountain-building events.

vard occur in flint nodules in the Gunflint formation in Michigan. They represent early plant remains, similar to algae or fungi, and have been
beautifully preserved in solid masses of silica. Age
measurements of the enclosing rocks indicate that
they are not less than 1600 million years old.
Plate IV shows what they look like under the
microscope.

Vast expanses of time were needed for the life
on the earth to grow more complex. For a billion
years it evolved from these primitive forms, leaving very little record of what was happening. The
time from 1600 million years ago to 700 million
years ago is almost a blank. Although the plants
and animals were developing into responsive, specialized, and well-adapted organisms, they had not
yet acquired the ability to grow a hard protective shell or structural hard parts that would be
preserved as a fossil record of their existence. But,
indirectly, the soft algal colonies of very early
times left a fossil record by their effect on the precipitation of calcium carbonate in sea water. This
was not sufficient, however, to permit detailed
studies of the organisms themselves.

About 600 million years ago the first important
fossil record started. Prominent types of organisms
found in a marine environment at this time are illustrated in Plate V.

We can best understand the streams and the
principal dates in the evolution of living things by
following them in the diagrams given in Fig. 19.
The search, the classification, the reconstruction
of the environment, as parts of the prodigious

114

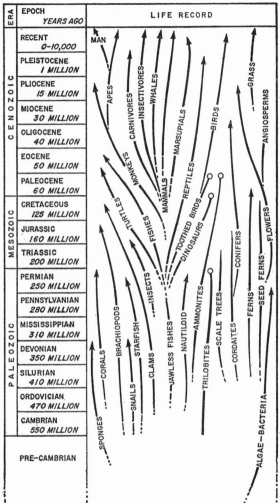

Fig. 19. Paleontology is the study of the evolution
of living things. This diagram summarizes the main
courses of development.

115

study that makes up the pages of the history of evolution, represent a major field in earth science known as paleontology.

The study of the fossil record of living organisms has been the basis of the study of geological sequences of events. Therefore, paleontology and historical geology are intimately interwoven. The vast areas of sedimentary rocks that cover much of the continental surfaces, that contain all the world's oil and natural gas and coal, and provide most of its soils for agriculture, are studied and mapped and classified into geological formations on the basis of their "geological age." The term "geological age" refers to a relative time scale not based on years ago but based on the established sequence of the evolution of plants and animals. Therefore, in order to establish the geological age of a section of sedimentary strata in any area, it is necessary to find fossils in these rocks and to recognize their position in the sequence. The geological age scale and the corresponding prominent living forms and evolutionary events are given in Fig. 19.

Strangely enough, it has been quite difficult to find out in actual years how long each part of the geological age scale is. We need to have crystals formed at some point in time in order to measure that time. The geological age scale primarily charts the time of deposition of sediments on the sea bottom, and this process does not commonly involve the growth of stable crystals. (There are some, however.) Therefore, the time measurements must be made with different rocks, such as have cooled

116

from hot magmas, that lie between the sedimentary sequences. Considerable uncertainty still exists, but the dates attached to the "geological ages" are probably correct within reasonable limits.

You will naturally wonder why the geological age scale ceases to list specific events earlier than about 600 my. ("my." stands for "million years ago") despite the fact that there were major geological events occurring in continental areas for 2000 million years prior to that. The reason is that the fossil record stops at 600 my., leaving the geologist no relative age scale to read. Therefore, the absolute measurement of geologic time by the methods described in Chapter IV must be brought into use to cover the 85 per cent of earth history for which fossils do not exist.

The Ice Ages

If it were possible to turn backward in time a short way—"short," that is, on the geologic scale —what would we see?

The most striking spectacle of all would be an extraordinary sea of barren ice spreading down from Greenland and the Arctic Ocean and gradually covering Canada and the northern part of the United States until it stood almost a mile thick with a front extending from New York to Oregon, as in Fig. 20. Thick forests of boreal spruce and pine covered the rest of the country to the south. Around the coastlines were broad strips of land, now under water; the level of the ocean had dropped 450 feet below present-day sea level.

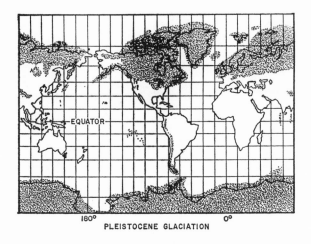

PLEISTOCENE GLACIATION

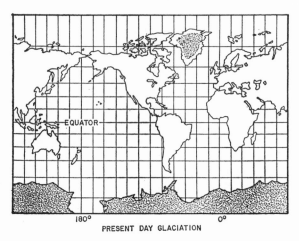

PRESENT DAY GLACIATION

Fig. 20. Approximately 20,000 years ago the Arctic ice cap extended with great thickness halfway down our continent. This was only "yesterday" in geologic time.

118

It is hard to believe, but this happened only 10,000 to 15,000 years ago, and it happened at the same time in the northern part of the Eastern Hemisphere. Climates the world over were greatly changed. Types of vegetation and animal life migrated to regions where they do not now exist. The ice is probably still receding today (see Fig. 20), and we see evidence that the sea level in general is still rising.

Going back further in time, we would find the ice receding and then advancing again several times. This continued until we get back to about 600,000 years; there is no evidence of continental glaciation from then until about 200 million years ago. The last three quarters of a million years, or rather this period of ice advance and recession, is referred to as the Pleistocene epoch, or familiarly as the "Ice Ages." It is a very recent time in earth history and is intimately linked with the earlier archaeological history of man.

The measurement of the time of these ice ages is an interesting demonstration of ingenuity in the face of difficult conditions. It has not been possible to use argon/potassium or similar radioactivity methods for this span of time, both because of the very small amounts of radiogenic elements formed compared to contaminations and because of the lack of formation of any crystalline substance that is caused by, or associated with, the ice advances. After a few preliminary remarks on the nature of the problem, let us follow the roundabout way in which this dating problem is being accomplished.

The evidence for the several ice advances appears in the form of planed surfaces of rock that once were hills, great piles of debris, such as boulders, sand, gravel, and wind-blown dust. As the ice overrode the land and receded again, it changed the drainage patterns by damming up valleys and basins, and formed the many lakes in the northern part of the continent, including the Great Lakes.

Glacial geologists have struggled with the history of these advances and have worked out four main advances which agree independently on the North American continent and in Europe. These are spaced by interglacial periods in which warm climates extended deep into the present polar regions. In these times the evolution of present-day man was rapid, so that the dating of the glacial events is intimately associated with anthropological dating and the unfolding of one of the most interesting studies in science—man's emergence to a position of dominance.

How is this dating accomplished? Rough estimates from the counting of annual layering in sedimentary clay deposits, thickness of soil cover, volcanic events, and evidence of man's advancing culture were assembled gradually but still were inadequate. It was not until three independent lines of study were brought together within the last few years that it has been possible to make much progress on this problem.

First, the study of ocean-bottom sediments was made possible by the development of coring devices that could be lowered to the ocean floors to obtain core samples of the bottom muds to depths of several tens of feet. The "piston corer," developed by B. Kullenberg for Hans Petterson's Swedish Deep-Sea Expedition of 1947, brought up cylindrical samples of the bottom material more than sixty feet in length, some of which represent a time span of more than a million years. These cores provide valuable records of recent geological history, as they are often undisturbed by the forces of erosion that act on continental materials. Of particular interest is the fact that the sediments contain the remains of single celled organisms called *foraminifera* which live in the column of ocean water above.

The various species of foraminifera live either on the deep ocean bottom or near the surface, and the skeletons of the different types can be selected from the oozes that make up the cores. These organisms have also evolved with time, and it is possible for paleontologists to use extinct species as a means of determining the geological period in which the organism lived. By studying living forms in different temperature environments, it has also been possible to distinguish the species that live in cold water on the bottom of the oceans from those that inhabit the warmer surface waters.

The next link in the chain of events leading to the answers we need was the development of a "thermometer" that would tell the temperature of the ocean water in the past. This seemingly impossible feat was accomplished by Harold C. Urey, a Nobel prize winner, and his associates at the University of Chicago. Urey predicted that in processes such as evaporation, precipitation, or biological secretion there would be a sufficient fractionation of the isotopes of elements of low mass to be detectable by very refined means of isotopic analysis. In other words, he found that when a glass of water evaporates, the three isotopes of oxygen (oxygen 16, 17, and 18) will not have the same relative abundance in the vapor as in the remaining liquid. Specifically, the lightest isotope, oxygen 16, will be more abundant in the vapor than in the remaining liquid. Similarly, the oxygen of the carbonate, or shell, would be slightly enriched in the heavier oxygen 18 isotope. Of particular interest, however, was the fact that the temperature of the water would affect the amount of the slight enrichment process. Urey predicted that a difference of one degree centigrade in water temperature would produce a difference of only .02 of 1 per cent in the ratio of oxygen 18 to oxygen 16 in the carbonate.

The extraordinary difficulty of detecting and measuring such small differences was finally overcome by redesigning the mass spectrometer to improve its sensitivity by a factor of ten. Finally, after intensive study and outstanding instrument design, it is now possible to measure relative temperatures in carbonate shell fragments to 1°C, if the fragments developed from water that contained the same ratio of oxygen 18 to oxygen 16. Actually, ocean waters differ considerably in this ratio, so that the job of measuring ancient ocean temperatures has been fraught with difficulties; it still is not often possible to place the temperature variations accurately on an absolute scale. But the "thermometer" is now reasonably well established and in many cases can be used for measuring past temperatures with a fair degree of reliability.

Finally, the development of the carbon 14 method of dating provided the last step leading to the solution of our problem. Although carbon 14 age measurement will not extend back in time beyond the last glacial advance, it has provided the yardstick for extrapolating time back through the entire glacial epoch.

Relying on the preliminary studies of Epstein, Craig, and others, an investigation by Cesare Emiliani and his associates at the University of Chicago has finally brought together all these developments in the following demonstration. It was reasonable to assume that the ice advances would be caused by, or reflected in, changes of temperatures of the main bodies of oceanic waters. How true this correlation would be in detail was to be discovered by using the carbonate "thermometer" of Urey on foraminifera of the kind that lived near the ocean surface.

It was first necessary to establish the relative temperature at which various species of foraminifera live at the present time. The species living within a few hundred feet of the surface of ocean waters were selected as giving best indications of detailed variations in climatic history, since the changes in deep ocean water are largely controlled by deep running currents bringing water from thousands of miles away. Then the isotopic ratios were determined in the same species in a number of cores taken thousands of miles apart. By measuring the same species of

123

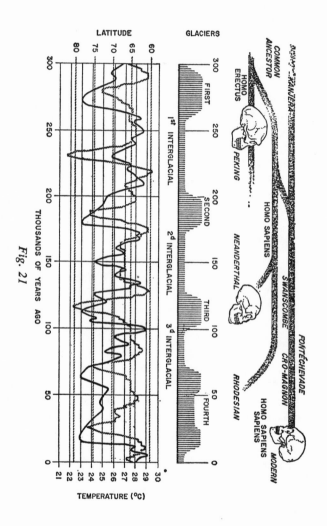

Fig. 21

124

foraminifera down the length of the core, representing progressively older periods of time, it was possible to observe the variation in temperature of the water of the ocean overlying that core in a series of climatic fluctuations. The picture of these climatic fluctuations appeared to be similar in cores taken from the Mediterranean, Caribbean, Equatorial Atlantic, and North Atlantic areas.

To establish a yardstick of time measurement in these investigations, Hans Suess and Myer Rubin of the Radiocarbon Laboratory of the U. S. Geological Survey determined the ages of the topmost foraminiferal remains by carbon 14 analysis and discovered that the most recent low point in ocean surface temperature came about 18,000 years ago. This agreed with other measurements as being the time of the maximum development of the last

---

Fig. 21. (opposite) The earth's climate in the last 300,000 years may have undergone the changes indicated in these two curves, which represent an attempt to date the Ice Ages. They correlate geological evidence of ice advance and recession with ocean temperatures and with calculations of variations in the sun's heat. The solid line shows the ocean temperatures derived from study of six cores obtained in deep-sea bottom borings. The gray line charts changes of heat received from the sun in northern latitudes. According to the calculations, latitude 65 receives as much heat from the sun in summer today as latitude 75 did 25,000 years ago. The similarity of the results of the two approaches to dating the Ice Ages is apparent in the curves. The skulls are associated with the epochs indicated on the chart. (After Cesare Emiliani)

glacial advance in North America. A composite curve showing Emiliani's findings in six deep-sea cores in Fig. 21 shows the climatic fluctuations during the glacial epoch.

There was rather extraordinary agreement between this composite curve and the predictions of the Serbian physicist, Milutin Milankovitch, who in the 1920s worked out the fluctuations in the reception of heat from the sun at various latitudes that might occur as a result of changes in the earth's orbit and in its axis of rotation. Emiliani's comparison of his oceanic temperature variations with these calculated variations in summer solar radiation received at latitude 65°N is also shown on the figure. The curve showing the variation in the sun's heat received in the northern latitudes is expressed as apparent shifts in latitude and shows an apparent 20° maximum equivalent latitude variation in a cycle of approximately 40,000 years.

Although this correlation is not yet proved, because direct radioactive age measurements cannot be extended beyond the most recent glacial advance, it is sufficiently good to constitute a provisional time scale for the entire glacial epoch and is now being tentatively accepted over other methods. If the correlation continues to hold in the future, even the fine details of the glacial advances will be known in the time scale, and a yardstick will be provided for the interpretation of the evolution of man from his predecessors.

No one knows for sure the cause of these ice advances, and we do not know whether another one is due in the next few tens of thousands of years. The balance between precipitation and evaporation in the polar regions is critically dependent upon atmospheric circulations and mean temperatures, and there are many possible ways in which this balance might be shifted.

Geologists have worked out the succession of ice advances and recessions independently on the North American continent and in Europe, and the results seem to agree reasonably well. Four main advances have been recognized, together with interglacial periods in which the warm climates extended even deeper into the polar regions than today. Fig. 21 shows a general pattern of these events.

Man's Early History

Man appeared on earth only yesterday in geologic time. More than 99.9 per cent of earth time had passed before the first evidences of the special aptitudes of man appeared in the fossil records. The particular study of man's development as a distinct species is part of the branch of science known as anthropology. The study of man's cultural history is known as archaeology.

Archaeological history, like geological history, has therefore been developed on a relative time scale, i.e., one based on something which has continuously changed with time. In geology we have seen that this changing "something" was the universal evolution of living organisms. In archaeology it has been the combined evolutionary changes

127

in man's physical frame and the nature and products of his culture.

It was a great day for archaeologists when W. F. Libby and his associates discovered and developed the use of radioactive carbon 14 for the measurement of very young ages. This method, described in a preceding chapter, is based on a 5600-year half-life so that it is particularly useful in the range of 500 to 30,000 years ago.

Nowadays carbon 14 measurements by the hundreds are illuminating man's cultural history. It could not have been more fortunate that carbon 14 happens to be formed by neutrons in the earth's atmosphere: carbon enters into so much of the matter connected with living things. Shells, flesh, hair, wood, peat—organic remains of all kinds—all contain carbon and can be used for age dating by this method.

For an example of the way in which an archaeological site reveals a carbon 14 record of man's cultural history and evolution, let us look at a recent excavation in the hills of northern Iraq. The site is the huge Shanidar Cave, now occupied by a clan of Kurdish goatherds and their animals. The exciting story unfolded in the course of the excavation has been told by Ralph S. Solecki, who initiated and supervised the work for the Iraqi Directorate-General of Antiquities and the Smithsonian Institution.

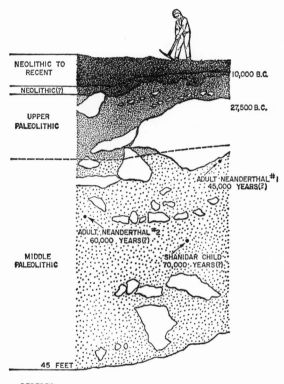

NEOLITHIC TO RECENT   10,000 B.C.

NEOLITHIC(?)

UPPER PALEOLITHIC   27,500 B.C.

ADULT NEANDERTHAL #1
45,000 YEARS(?)

ADULT NEANDERTHAL #2
60,000 YEARS(?)

SHANIDAR CHILD
70,000 YEARS(?)

MIDDLE PALEOLITHIC

45 FEET

BEDROCK

Fig. 22. Archaeological excavations in the great Shanidar Cave in northern Iraq show a record of late and
early man going back about 100,000 years. The record has been dated by carbon 14 in the upper layers.
(After Dr. Ralph Solecki)

129

Carbon 14 and a Cave

The cave had apparently been inhabited by man for about 100,000 years, and the layer of material 45 feet thick on the floor reveals a history of the occupants going back to the predecessors of the modern species of man. Dr. Solecki describes the excavation as passing through four main layers, with different soil color and artifacts, as indicated in Fig. 22.

The top layer is a greasy soil dating from the present back to some time in the New Stone Age, about 7000 years ago. In it are ash beds of communal fires, bones of domesticated animals, and domestic tools such as stone mortars.

The next layer dates back to the Middle Stone Age, about 12,000 years ago, according to carbon 14 measurements. It contains beautifully chipped projectile points, bone awls for sewing, and engraved pieces of slate suggesting early art work. But conspicuously lacking is any evidence of animal domestication, agriculture, or pottery making. Heaps of snail shells give a suggestion of the diet.

The third layer dates from about 29,000 to more than 34,000 years ago, according to radiocarbon measurements on charcoal. (There is no record in the interval between 17,000 and 29,000 years ago.) This culture is Old Stone Age. Many flint woodworking tools and scraping tools were found, similar to others in the late Paleolithic culture in Europe.

In these three top layers the peoples were all

130

presumably Homo sapiens. In the bottom layer, extending from a depth of 16 feet to bedrock at 45 feet, were found the remains of extinct Homo neanderthalensis, or Neanderthal man. Not only were his crude tools discovered but also three skeletons.

In the bottom layer there is an 8-foot section with an especially heavy concentration of remains of fires, suggesting continuous occupation of the cave through a cold period. There also is a layer of stalagmitic lime, proof that the period was un- usually wet—the only one in the history of the cave. The bottom layer may represent time back to 100,000 years ago.

Unfortunately, the radiocarbon of those times is so nearly gone that it can no longer be used effectively. Although it may eventually be possible to extend the method by enrichment of the carbon isotope before measurement, problems of back- ground radiation and of contamination have been insuperable. Dating of most of the Pleistocene and the period of evolution of modern man is currently beyond the reach of the technique. Thus, at pres- ent, the dating must be done by geological cor- relations.

This illustration of the record of man's progress again shows the painfully slow changes in early culture as millennium after millennium passed without variation in rudimentary tools, followed by the almost sudden burst of development that occurred during the last advance and recession of the ice.

Correlation of all present-day knowledge of the

131

evolution of Man with Emiliani's time scale suggests the progression chronologically charted in this table:

Neanderthal man	Became extinct about 50,000 years ago
Fontéchavade remains	About 100,000 years ago
Swanscombe skull bones	About 125,000 years ago
Pithecanthropus, Sinanthropus, and Atlanthropus	About 200,000 years ago
South African man-apes	200,000 to 400,000 years ago

Thus, if Emiliani is correct, the evolution from Pithecanthropus to Swanscombe man took about 3000 generations, and from Swanscombe to Modern man, only about 1000. The unsuccessful Neanderthal man lasted only about 2000 generations.

The development of man's particular faculties was rapid in this last glacial period. Changes of environment were severe, and his existence depended upon rapid adaptability to climate, food supply, and defense. He survived through craftiness, taking advantage of natural aids wherever possible. Constructing shelters against climate, developing weapons and utensils for food and defense, using fire for his comfort, he gradually became less the nomadic predator and more a member of tribal groups with community of interest. This directed him to the beginnings of agriculture

132

and domestication of animals, and the start of modern civilization.

The use of fire goes far back in man's history. Certainly Neanderthal man used it. Probably the earliest date that can be safely assumed for the use of fire is that of Sinanthropus, or Peking man, the primitive human being who had a brain only two thirds the size of modern man's. Peking man, according to geological estimates, lived about 250,000 years ago. The startling discovery in South Africa a few years ago of Australopithecus prometheus, in deposits showing traces of burned bone, suggested the use of fire by a sub-human creature with a man-ape cranium, but there is insufficient supporting evidence to establish this as fact.

How badly we need an age-dating method that bridges the gap between carbon 14 and the conventional geological methods! The major problem is lack of materials that were formed at the time of the event and preserved intact ever since. Unfortunately, the best material, carbon in the charred wood, which is so commonly found in archaeological sites and which is so useful in carbon 14 dating, is no longer of use in these earlier times.

The answer may have to come from a complete dating of past climates and ice ages and better geological correlation with these.

CHAPTER VI

The Earth's Beginnings

We cannot consider the origin of the earth without bringing in the origin of the solar system. Because the solar system contains a wide range of elements, we must find some way to account for these, and in so doing we should consider other stars besides our sun. It seems probable that many, if not all, the elements except hydrogen are produced in the interior of stars by thermonuclear reactions. These reactions require exceedingly high temperatures, which, according to calculations, exist only in the interiors of certain stars at a particular time in their history. (Estimates of the temperatures needed are charted in Fig. 23.)

In the center of the sun the pressure is so high that ordinary elements would not be able to maintain the electron shells we are familiar with on the surface of the earth. If its temperature were not exceedingly high, the sun would be a much smaller and denser body than it is known to be.

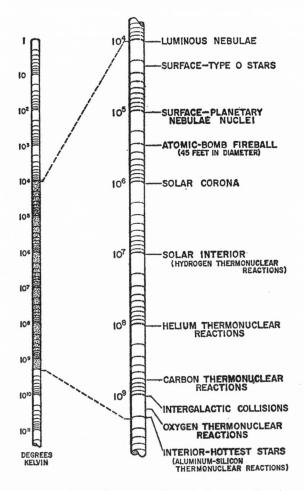

I
10
10^2
10^3
10^4
10^5
10^6
10^7
10^8
10^9
10^{10}
10^{11}

DEGREES
KELVIN

10^4 ——LUMINOUS NEBULAE

——SURFACE—TYPE O STARS

10^5 ——SURFACE—PLANETARY
NEBULAE NUCLEI

——ATOMIC-BOMB FIREBALL
(45 FEET IN DIAMETER)

10^6 ——SOLAR CORONA

10^7 ——SOLAR INTERIOR
(HYDROGEN THERMONUCLEAR
REACTIONS)

10^8 ——HELIUM THERMONUCLEAR
REACTIONS

——CARBON THERMONUCLEAR
REACTIONS

10^9 ——INTERGALACTIC COLLISIONS

——OXYGEN THERMONUCLEAR
REACTIONS

——INTERIOR—HOTTEST STARS
(ALUMINUM-SILICON
THERMONUCLEAR REACTIONS)

Fig. 23. A popular current hypothesis states that the elements other than hydrogen in the universe were formed by thermonuclear reactions in the center of hot stars. Estimated temperatures for such reactions exceed 10^9 degrees C.

135

Thus, a balance between the temperature and the mean density of the sun must exist. The sun loses much energy in radiation from its surface, so there must be a continuous transfer of energy from the interior to the surface. If this energy came only from cooling of an initially hot body, the sun would shrink in size as its interior temperature dropped. In the shrinking the system's gravitational energy would diminish just as it would in the fall of any body toward the earth. When a body falls toward the earth under gravity, work is being done to give it an acceleration. It acquires kinetic energy at the expense of the total gravitational energy of the system. The gravitational energy release in the shrinking would be fairly quickly used up by the radiation loss at the surface. Knowledge of the minimum age, density, composition, and other characteristics of the sun does not permit the conclusion that its energy is derived solely from gravity in a process of slow collapse. Calculations along these lines were initially responsible for the prediction that nuclear energy was a crucial factor.

Nuclear Fires of the Sun

The sun is composed predominantly of hydrogen. We now know that at the temperatures that must exist in the interior of the sun (Fig. 24) to maintain its relatively low density, hydrogen nuclei will react with each other and form helium, with a great release of energy. In other words, hydrogen is being burned up leaving helium as an "ash." The

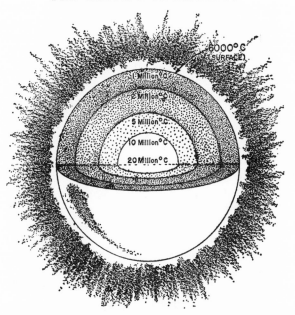

Fig. 24. The temperature distribution and energy production in the sun have been a subject of much interest, as they were important in predictions relating to thermonuclear reactions. Present estimates of temperature are shown above.

temperature of this transformation is of the order 10^7 degrees centigrade.

Why does not the hydrogen react explosively and blow the sun apart, as in a hydrogen-bomb detonation? The answer is that the reaction is automatically controlled by expansion of the sun. The release of thermonuclear energy raises the internal pressure and causes the sun to expand, thereby

137

reducing the probability of nuclear interactions. The two effects are in balance.

But it can easily be seen that as the helium builds up it will also slow down the hydrogen reactions. For the same energy release there must be a higher temperature to increase the probability of reactions among those atoms not yet converted to helium. Thus the interior temperature of the sun is probably slowly increasing. This slow increase of temperature eventually will cause the helium to react, producing carbon 12, oxygen 16, and neon 20.

The same process of burning up the lighter elements, accompanied by increase of temperature, continues through the evolution of a star. Starting with almost a pure hydrogen composition and an internal temperature of about 12–15 million degrees, the star burns its hydrogen and manufactures helium with a gradually increasing temperature. The temperature increases to about 100 million degrees by the time the hydrogen has half gone and the helium starts to burn. Just as the hydrogen becomes exhausted, so eventually does the helium. By the time the temperature has risen to about 600 million degrees, carbon is being converted to sodium, magnesium, and neon. Further increases in temperature cause reactions that produce aluminum, silicon, sulphur, phosphorous, chlorine, argon, potassium, and calcium. Finally, at about 2000 million degrees these elements are converted into what is known as the iron group—titanium, chromium, manganese, iron, cobalt, nickel, copper, and zinc.

Probably the entire star would not become converted to these elements because the temperature would decrease toward the surface. How the elements, from hydrogen to the heaviest, are distributed between the surface and center is difficult to estimate, but in the sun it appears likely that at least the outer half of the radius is made up of fairly homogeneous material.

In this process of evolution, well summarized recently by the British mathematician Fred Hoyle, William A. Fowler, of California Institute of Technology, and others, the star's interior temperature will increase and the star will become extremely expanded and more luminous. Also its surface temperature will increase. It is calculated that the sun, which is fairly youthful in this stellar life span, eventually will become 1000 times more luminous and have a radius 100 times greater than at present. Finally, it will shrink to one twentieth of its present size as the nuclear fires within die out, and it will become the type of star known as a "white dwarf."

This hypothesis of stellar evolution is supported by observations of the relationship between luminosity (brightness) and surface temperature (color) of numbers of stars. Fig. 25 shows, for example, a study made by H. C. Arp and A. R. Sandage of the Mount Wilson and Mount Palomar observatories, in which this relationship is plotted for a number of stars belonging to the globular cluster Messier 3. All the stars plotted, it is believed, have a mass fairly similar to that of the sun.

The plot makes a definite pattern. The hypothesis holds that a star will start its evolution close to

139

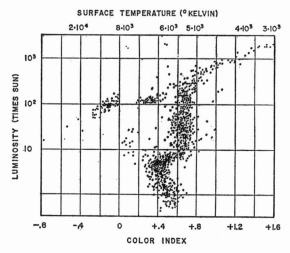

Fig. 25. The evolution of stars is an exciting new subject. An important piece of evidence in a chain of reasoning is this compilation of the relationship between color and brightness in a number of stars in a globular cluster. (After H. C. Arp and A. R. Sandage of the Mt. Wilson and Mt. Palomar observatories)

the bottom. Its luminosity (dependent on size and temperature) increases, until it reaches the "giant" branch at the upper right. Finally, the luminosity decreases slightly, but the surface temperature increases greatly, and the star follows the trend shown in the horizontal branch at the upper left. In this stage the light that is emitted shifts farther to the ultraviolet (indicating a higher surface temperature). This ultraviolet is strongly absorbed by interstellar gas and our own atmosphere, and in time the star disappears from our view.

Origin of the Elements

After a brief look at these hypotheses, let us now return to the origin of the elements that make up our solar system. It is not yet known for certain whether stars considerably larger than our sun might be capable of making the heavier elements, but it is an attractive thought. If true, it seems likely that the elements would have been dispersed into space within our galaxy through the explosions of supernovae. These exploding stars have been estimated to have been frequent enough to account for the small amount of material heavier than helium in the basic composition of our galaxy.

One of the problems in this hypothesis is the apparent uniformity of composition of the universe. The American astronomer Armin J. Deutsch has discussed the evidence from solar, stellar, and interstellar absorption spectra (the analysis of light from the sun and stars), indicating the extreme uniformity of their composition if adequate corrections are made for temperature and other effects. The observed "cosmic" abundances of the elements are indicated in Fig. 26.

Hydrogen greatly predominates, with helium next, and all the rest of the elements make up less than 1 per cent of the total. We are very interested in the relative abundances of the elements heavier than helium, because, as we shall see from the relationship between uranium and lead isotopes on the earth and in meteorites, the elements in this part of our galaxy are no older than a few billion years.

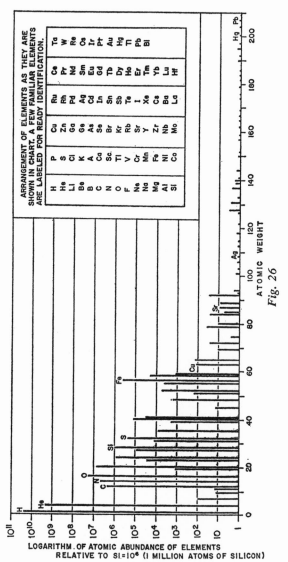

ARRANGEMENT OF ELEMENTS AS THEY ARE
SHOWN IN CHART. A FEW FAMILIAR ELEMENTS
ARE LABELED FOR READY IDENTIFICATION.

Fig. 26

ATOMIC WEIGHT

LOGARITHM OF ATOMIC ABUNDANCE OF ELEMENTS
RELATIVE TO Si=10⁶ (1 MILLION ATOMS OF SILICON)

142

Does this mean that the entire universe is only a few billion years old? Or just our galaxy?

Turning to our planetary system, we have the following evidences and problems. The densities (calculated for zero pressure) compared with the size of the bodies near the sun are as follows:

	Mass (compared to earth)	Calculated Mean Density (zero pressure)
Moon	0.012	3.3
Mercury	0.054	?
Mars	0.108	3.6 – 4.0
Venus	0.81	4.3 – 4.7
Earth	1.0	4.4

This suggests that the larger the body, the heavier is the material it is made of. But is this due to a greater abundance of heavier elements brought together at the start? It is difficult to believe that the moon and earth were not originally of the same composition, since they presumably condensed from the same cloud of gas and dust.

It has often been stated that the earth's core of iron and the moon's lack of it cause the difference.

Fig. 26. (opposite) If the elements were formed in thermonuclear reactions at the center of hot stars, their relative abundances should be predictable to some extent. For this we have to know the neutron capture cross sections (the probability of interaction between a given nucleus and an incident neutron) and the stability of the various nuclei. It has been of interest to correlate the observed abundances in nature with the predicted ones. The agreement is quite good.

But why would the abundance of iron in the earth differ from that in the moon? G. J. F. MacDonald has recently proposed that the relative abundance of the heavier elements in the planetary bodies was initially the same. In the earth, however, the pressure was not sufficient to strip electrons from atoms, as in the sun, but was sufficient to result in a separation of oxygen from iron and possibly silicon in the central region. The lower melting temperature of the pure metal phases permitted fusion.

It is still bothersome, however, that the ratio of iron to silicon in the sun appears to be much less than in the earth or meteorites. As the temperature of the interior of the sun is definitely not sufficient to be manufacturing silicon, magnesium, or iron, the sun inherited its supply of these elements from a source, presumably (but not necessarily) the same as that of the planetary bodies. This source probably developed early in the history of our galaxy as a cloud of dust or gas of cosmic composition, and was one of a large number of such clouds condensing into stars similar to the sun.

From a Cloud of Dust

In the last few years G. P. Kuiper, developing and modifying the ideas of C. F. von Weizäcker, has contributed much to an hypothesis of wide acceptability on the origin of the solar system. The central part of the cloud condensed to form the sun. The rotating disc of dust and gas which formed quickly from the nebula surrounding the

144

sun broke up into eddies of irregular size and arrangement within any radial zone, but the size of the eddies increased generally toward the outer zones. The eddies in any zone coalesced as they met each other, until finally there remained separate masses of gas at different radii from the sun. These masses of gas and dust have been referred to as "protoplanets."

The theory appears to provide rather nicely for the spacing and total mass of the protoplanets, the distribution of angular momentum in the solar system, and the sense of the rotation of the separate masses.

Among astronomers the formation of the planets from the protoplanets is a lively question at present. Much evidence must be accounted for. In mass the terrestrial planets are very much smaller than the protoplanets were, and very different in composition. The protoearth was a low-temperature gaseous mass about 500 times the planetary mass; the sun had not condensed sufficiently to be hot. The protoearth gas cloud was composed of hydrogen, helium, neon, methane, ammonia, and, possibly, some water vapor. Materials of which the present earth is made were condensed and eventually spiraled into the center. A secondary condensation nucleus formed the moon.

When the sun contracted and became bright, its radiations energized the cold gases of the protoplanets. Individual particles acquired velocities that permitted them to escape from the protoplanet's gravitational field. The escape of these gases, which amounted to 99 per cent of the protoplanet mass,

145

was essentially complete, even of the heavier ones like krypton and xenon. According to Kuiper's estimates, the time required for absorbed solar radiation to cause this removal of gases was probably several hundred million years.

We can follow several lines of evidence in an attempt to form an hypothesis for the early stages of the accretion and development of the earth system. The surface of the moon, which has no atmosphere and has retained its original character since its beginnings, shows great craters and scars, as if struck by large objects in its growth (see Plate VI). The origin of the iron and stony meteorites must be accounted for, as well as the apparent difference in iron content of the earth and moon.

The total amount of energy released by the infall of material to form the earth was not sufficient to have raised its temperature to its probable present value. Continuous radiative loss of heat outward also reduced the retained energy. So, depending on the rate at which the earth accreted, it could have been formed at various temperatures probably not exceeding about 2000° C.

Harold C. Urey, basing his arguments on the lack of concentration of volatile elements at the surface, has suggested that the temperature of the earth's surface in its final stages of accumulation probably was not higher than 200° C. Earlier theories, which conflicted with each other, were largely based on the assumption that the earth was molten in the beginning. Today the general belief is that initially the earth was fairly cold. Radioactivity, then 15 times more intense than it is now, and energy that may have been released when the iron core material settled to the center raised the temperature to a level at which the earth was essentially molten. It was at this time that the gravitational separation of core, mantle, and probably some of the materials of the ocean and atmosphere occurred.

Sir Harold Jeffries has developed the theory of a cooling molten mantle, crystallizing from the base upward, with heat lost first by convection in the liquid and later by conduction in the solid. His outstanding treatise has encountered the objection that the earth probably formed as a solid body, but the theory could apply if the mantle ever became molten at a later time. The crystallization from the base outward would result from the fact that a mixture of the oxides of the rock-forming elements solidifies over a range of temperature; the more stable forms would crystallize at a higher temperature and sink in liquid of lower density.

The Answer

Let us now return to the measurement of the age of the earth and solar system. So far, we have discussed age measurements that have depended upon parent-daughter relationships in a mineral that contains the parent element. Another interesting part of the field of age measurement relates to the slow change of the isotopic abundance in elements in which one or more isotopes are radioactive and breaking down, or radiogenic and building up.

147

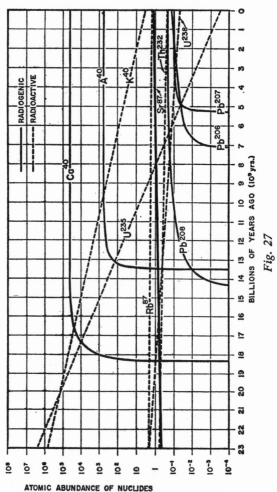

Fig. 27

148

If we know the half-life of a radioactive isotope, we can calculate the change of its abundance relative to the stable isotopes of the same element and see what happens as we go back in time. In Fig. 27 curves are shown for the changing abundances of potassium 40, rubidium 87, uranium 235 and uranium 238, and thorium 232.

As an example of radiogenic isotopes building up with time, let us look at the element lead. Lead 204 has no known long-lived parent so, presumably, has not changed in abundance significantly in the last few billion years. Lead 206, 207, and 208 are increasing at different rates because they are formed from the breakdown of U^{238}, U^{235}, and Th^{232}, respectively. If we now calculate the change in the ratios 206, 207, and 208 with time going backward, we get curves as shown in Fig. 27.

In these curves we have assumed average estimated values for the abundances of uranium, tho-

Fig. 27. (opposite) Calcium 40, argon 40, strontium 87, lead 206, 207, and 208 are examples of radiogenic nuclides. They are increasing in abundance with time because they are the decay products of potassium 40, rubidium 87, uranium 238, and thorium 232. This means that, going backward in time, there would be less of them until ultimately there was zero. Because the parent element is decaying exponentially, the drop-off of the daughter to zero appears to be quite a sudden phenomenon on the chart. These zero values set the maximum age for the origin of these pairs of elements. Lead 207 has the lowest maximum age, and this sets the limit for our part of the universe.

rium, and lead. Other values would change the time scale on these curves but not their shape. In each case there is some limiting time in the past beyond which we cannot go because the radiogenic isotope would be reduced to zero.

For example, in the figure we see that lead 207 is reduced to zero at about 7 billion years ago, which suggests that the association of uranium and lead in at least our solar system could not have existed earlier. As it seems likely that these heavy elements were all created together, the maximum time for the creation of the elements appears to be about 7 billion years ago.

Astrophysical calculations set an age limit for the galaxy's globular clusters at about 6.5 billion years. Being almost certainly younger than these, the sun is estimated to have an upper age limit of about 6 billion years.

Now let us approach the problem from the opposite direction and examine the oldest available material to establish a limit on how *young* the earth could be. The oldest measured rocks are 2.8 billion years old, and we know the earth is older than that. Argon/potassium and strontium/rubidium measurements on stony meteorites show a good maximum age grouping at 4.5 billion years. These objects are made of minerals which could not have been formed under pressures such as exist in our earth's deeper regions, so they must have been derived from the collision of small bodies or else were recrystallized from some other kinds of minerals or represent parts of the near-surface layers of a larger body. Thus, the age of 4.5 billion years

150

simply represents the time at which an event happened, which probably was similar to the separation of the earth's core and mantle.

Using a combination of the lead isotope ratios in iron meteorites and the curves shown in Fig. 27 based on the U/Pb ratio in the earth's upper mantle, we also get an age of 4.5 billion years, which suggests that the major separation of the earth into iron core and stony mantle occurred at approximately the same time as the development of meteorites.

Going back to our thoughts on the early history of the earth, we decided that a considerable length of time was needed for radioactivity to heat the earth from an initially cool beginning. Thus, the commonly quoted age of 4.5 billion years for the earth is a lower limit only, and the true age is probably hundreds of millions of years greater. As you will recall, it probably took several hundred million years to remove the gases from the protoplanets. If we add this period to the time needed to heat the earth enough for separation of mantle and core and the time needed for the sun to condense and become hot, we account for the period between 4.5 billion years ago and 6 billion years ago, the estimated time of the sun's origin.

So, in summary, we see that the elements in our part of the universe, the globular clusters and probably our galaxy formed 6.5 to 7 billion years ago; the sun condensed 6 billion years ago; the protoplanets reduced to planets about 5 billion years ago; the chemical separations within the terrestrial planets and parent bodies of the meteorites oc-

151

curred 4.5 billion years ago; and formation of a
lasting earth crust started 2.8 billion years ago.

How majestic are these broad reaches of time!
Looking into an abyss, one senses the gigantic
form of the void only in comparison to one's own
minute stature. It is almost incomprehensible that
only a few billion years ago our galaxy was born
in a giant bomb-flash of nuclear energy. What an
inspiring picture of the process of creation! But
awesome and inspiring as it is to contemplate this
mighty spectacle, the true reward is not to be found
in whether our calculations are correct, give or take
a few million years; it lies in the discoveries, in the
advancement of human knowledge and philosophy
that are the inevitable products of scientific search
for law in nature.

BIBLIOGRAPHY

1. *Physical Geology* by L. Don Leet and Sheldon Judson, Prentice-Hall, New York
   A good general textbook on physical geology.

2. *The Scientific American* book series, Simon and Schuster, New York
   Represents compilations of articles in this magazine on specific topics. In particular, the volumes: *The Planet Earth, The New Astronomy,* and *The Universe.*

3. *The Earth and Its Atmosphere,* edited by D. R. Bates, Basic Books, Inc., New York
   Contains articles by fifteen scientists on the crust and interior of the earth, its origin and ultimate fate; on the oceans, atmosphere, ionosphere, magnetic storms and cosmic rays.

4. *Nuclear Geology,* edited by H. Faul, John Wiley and Sons, New York
   A symposium on nuclear phenomena in the Earth Sciences, including age measurement, earth heat and radioactivity, instruments and uranium prospecting techniques.

153

INDEX

Absorption spectra, 141
Actinium series, 86
Advisory Committee on Biology and Medicine, report, 43
Age, of earth, 151 f.
of galaxy, 143, 150 f.
of sun, 150 f.
of universe, 110, 143
Age of the Earth, 88
Age measurement, history of, 85 ff.
by radioactivity, 76 ff.
Agriculture, beginnings of, 132 f.
Aldrich, L. T., 92
Alpha particles, 46
and heat, 57 ff.
Anthropology, defined, 127
Archaeological history, 127 ff.
Archaeology, defined, 127
Argon, 81, 85, 103, 119, 149, 150
in age measurement, 92 ff.
Argon-potassium method, 48
Aristotle, 22
Arnold, James R., 107
Arp, H. C., 139
Atmosphere, 18, 82
Atomic Energy Commission, 42
Atoms, binding force of, 45 f.
Australopithecus prometheus, 133
Avogadro, Amadeo, 95
Avogadro's Number, 95, 99

Barghoorn, Elso S., 113

Basement complex, 73
Becquerel, Henri, 86
Beta particles, 47
in generation of heat, 58
Binding force of atoms, 45 f.
Bullen, K. E., 28

Caesium 137 in fallout, 40
Calcium, affinity for Strontium 90, 43
California Institute of Technology, 17
Carbon 14, in age measurement, 91, 105 ff., 123, 125, 128, 133
half-life of, 106
Carnegie Institution, 90
Centrifugal force of earth rotation, 15, 27
Civilization, beginning of, 133
Colorado Plateau, 31
Colorado River, 73 f.
Compression, effect on earth's crust, 75
Compression waves, in earthquakes, 25
Compton, A. H., 50
Compton scattering effect defined, 50, 51
Concordancy in age measurement, 89, 105
Contamination, correction of, 84 f., 92, 97, 109
Continental masses, growth of, 38, 113
stability of, 31, 37
Convection and conduction of heat, 61 ff., 70
Cosmic radiation, 106
correction for, 108

155

Mohorovicic discontinuity,
18, 26, 34 f., 37, 64
Moon, the, 71, 143 f., 146
Mountain belts, 75 f., 88,
113
Mountain making, 31 ff.,
36 ff., 70, 72 ff., 87,
113
"My.," definition of, 117

Neanderthal man, 131 ff.
use of fire by, 133
Neutrino, 47, 48
Neutrons, 45
New Stone Age, 130
Nier, A. O., 89
Nuclear fission, discovery
of, 89
Nucleus, defined, 45
Nuclides, atomic abun-
dance of, 148
radioactive, rate of de-
cay, 82

Ocean area, ratio to land
area, 31
Ocean temperatures, 122 ff.
Oldham, R. D., 17
Old Stone Age, 130
Orogeny, 72
Ovid, 22
Oxygen, as component of
rocks, 20 f.
isotopes of, 46, 122

Paleolithic culture, 130
Paleontology, 31, 115 f.
Peking man, 133
Petroleum, 73
Photon, defined, 51
"Physics in the Earth"
series, 88
Piggott, Charles S., 90

Planets' mass and density,
143
Pleistocene Age, 119, 131
Poles, the, 15
Potassium, breakdown of,
99 ff.
in granite and volcanic
rocks, 64
migrational qualities,
66 f.
use in age measurement,
80, 91
Powell, John Wesley, 31 f.
Pre-Cambrian period, 75
Primary waves in earth-
quakes, 24
Protons, 45
"Protoplanets," 144 f.

Quantum defined, 51

Rad, definition of, 39
Radiation, cosmic, 106,
108
dosages, limits, 42 f.
Radioactive breakdown,
46 f., 78, 100 ff.
related to isotope abun-
dance, 81
Radioactive components,
migration of, 69
Radioactivity, 39 ff.
artificially produced, 89
contamination of fish by,
42
fallout, 40 ff.
injurious sources of,
40 f.
reader bombarded by, 39
value of energy from,
70 ff.
value of heat from, 60
Radioactivity in Geology,
87

158

159

DATE DUE

WORLD INDUSTRY STUDIES

Edited by Professor Ingo Walter,
Graduate School of Business Administration,
New York University

Politics vs Economics in World Steel Trade

Kent Jones

Assistant Professor of Economics
Babson College, Massachusetts

London
ALLEN & UNWIN
Boston   Sydney

Allen & Unwin (Publishers) Ltd,
40 Museum Street, London WC1A 1LU, UK

Allen & Unwin (Publishers) Ltd,
Park Lane, Hemel Hempstead, Herts HP2 4TE, UK

Allen & Unwin, Inc.,
8 Winchester Place, Winchester, Mass. 01890, USA

Allen & Unwin (Australia) Ltd,
8 Napier Street, North Sydney, NSW 2060, Australia

First published in 1986

British Library Cataloguing in Publication Data

Jones, Kent
    Politics vs economics in world steel trade.——
(World industry series; 4)
1. Steel industry and trade
I. Title   II.Series
338.4'7669142   HD95105
ISBN 0−04−338118−9

Library of Congress Cataloging-in-Publication Data

Jones, Kent Albert.
    Politics vs economics in world steel trade.
(World industry series; 4)
Bibliography: p.
Includes index.
1. Steel industry and trade——Government policy.
2. Free trade and protection——Protection. I. Title.
II. series.
HD9510.6.J66   1985   382'.45669142   85—13545
ISBN 0−04−338118—9 (alk. paper)

Set in 10 on 11 point Times by A. J. Latham Limited, Dunstable, Beds.
and printed and bound in Great Britain by
Biddles Ltd, Guildford and King's Lynn

Contents

Editor's Foreword

In many ways, steel is an ideal subject for this series of WORLD INDUSTRY STUDIES. It is by now an old industry, dating back well over a century, with well-known and internationally-available production and product technologies. It is an industry of great national importance for reasons of industrial output and employment, regional concentration, structural linkages to other sectors of the economy, and defense preparedness, among others. And it is an industry that has been subjected to great competitive stress the world over. Traditional players have been placed under enormous pressure and many fall by the wayside. New players have emerged, armed not only with low raw-material and labor costs, but also with state-of-the-art technology. Patterns of international trade, both in terms of the volume and direction of steel shipments, reflect these essentially market-driven forces.

Against this dramatic backdrop of competitive shifts in the world steel industry, most of which are not difficult to predict using conventional economic analysis, a broad range of business and public-policy decisions have had to be made, both strategic and tactical. At the corporate level, these decisions have focused on the ability to meet competitive threats and opportunities, the need perhaps to disengage from or focus more heavily on the steel business, the need to source certain products offshore or to target certain foreign markets, and the likelihood of obtaining government aid through subsidies or protection against imports. At the political level, the critical decisions focus on the adjustment costs associated with competitive shocks, the relationships of these costs to a range of other economic considerations and to political commitments made to constituents, as well as the pros and cons of protectionism in the steel sector and its broader implications.

In all of these dimensions, there has been a great deal of action in recent years among the four principal poles of competition in the global steel industy — Western Europe, the United States, Japan and major developing country suppliers like South Korea, Taiwan, Brazil and Mexico — as well as lesser suppliers like South Africa and Eastern Europe. The United States has been highly protectionist with respect to this sector. In Western Europe the steel industry epitomizes struc-

tural rigidities and the lack of political will to adapt to market forces. Japan continues to have a sectoral advantage in steel and is probably hardest-hit by protectionism elsewhere in the world, but even here the handwriting is on the wall and restructuring is inevitable. And the developing countries must search for market openings as best they can, regularly facing decisions on how many resources to allocate to steel, and on what terms.

These conditions will continue for the foreseeable future. Steel remains a critical industry, with large labor forces as well as dedicated and regionally concentrated facilities rendering it all the more sensitive from a political as well as economic standpoint. The present volume does a superb job of carefully and dispassionately analyzing these issues using the tools of modern political economy. The lessons learned here will carry well beyond the global limits of the steel industry.

Author's Preface

Among the far-reaching changes that have occurred in the structure of world industry in the postwar period, the shift in the pattern of comparative advantage in steelmaking has been particularly dramatic. The once powerful American steel industry, which at the end of the war appeared impregnable from foreign competition in its own market, began losing market shares to imports in the late 1950s and in 1984 yielded 26 per cent of its domestic market to foreign producers. European steelmakers, who had rebuilt their mills after the war and by the 1960s had regained a large measure of their prewar prominence, none the less began to suffer from an overbuilt industry and declining worldwide competitiveness in the 1970s. By 1980, market conditions deteriorated to the point where the Commission of the European Community declared a state of 'manifest crisis,' requiring steel producers to submit to mandatory production quotas and minimum prices. Many factors had contributed to the deteriorating position of the American and European industries, including technological developments, changes in the structure of steel demand, and long-term international shifts in the relative cost of inputs. It was in any case becoming clear that comparative advantage in steelmaking was moving towards the relative newcomers in steel production – Japan and some of the newly industrializing countries. In the United States and Europe, the traditional steelmaking areas of the world, the industry most closely linked with economic power and growth in the nineteenth century was approaching the end of the twentieth century in a state of perpetual crisis and decline.

Yet the inevitable and inexorable process of industrial restructuring and adjustment that accompanies any basic shift in underlying market forces has met with remarkable resistance in the case of steel. The painful signs of adjustment in such a visible and basic industry have struck at the national consciousness, raising questions of the very well-being and vitality of the national economy as a whole. Thus the closing of a major steel plant in Lackawanna, New York or the French government's decision to reduce its nationalized steel industry's capacity by 25 per cent – followed by rioting among steel workers in Alsace-Lorraine – become more than sensational front-page bad

news; they generate calls for governments to 'do something' to prevent the extinction of an industry symbolic of economic health and growth.

The clash of economic forces with national pride and deep-seated fears and suspicions of foreign competitors on the international marketplace provides the formula for bitter and self-defeating international policy disputes. In the case of steel, this conflict has emerged as the growth of trade protectionism. The goal of this study is to identify the historical, economic and political forces that have shaped the current pattern of steel trade policy and the environment for economic adjustment in the United States and the European Community. In pursuing these goals, the study also sets out to develop a framework for understanding the factors that give rise to protectionism and cause it to spread worldwide. The steel industry provides an important case study of market forces placing severe pressure on firms to undertake painful adjustment measures — including layoffs, plant closings, and expensive modernization programs — and of the ability of the industry to resist those market forces through political influence.

The study begins with a perspective on the underlying economic and political forces that motivate steel trade policy (Chapter 1). While economic arguments for protecting the domestic steel industry from imports are tenuous at best, the association of the industry with economic well-being creates strong political support for such policies. Chapter 2 chronicles the development of steel trade policy and its legacy of protectionism since the late nineteenth century, providing historical background and precedents for current steel problems. In Chapter 3, a general framework for understanding steel protectionism is laid out, focusing on the inherent difficulties in the steel industry of adjusting to market changes and the channels of influence that it can exploit to achieve political influence and resist the adjustment process.

Steel protectionism in the United States and Europe can therefore be understood as a deeply-rooted process of resistance to changes in international market forces. Chapter 4 examines the long-term trends in economic factors determining international steelmaking competitiveness and their impact on steel producers in the United States and the European Community. Surges in steel imports emerge as more a symptom than a cause of the underlying shift in comparative advantage. Chapter 4 also examines the available evidence on the impact of government involvement on the long-term competitive position of national steel industries. While the facts do not seem to support the contention that government involvement has supplanted market forces in determining trade flows, it is clear that governments' role in steel production and trade has poisoned trade relations, thus to a certain extent choking off steel trade flows. Chapter 5 traces the course of United States and EC policies towards their national steel industries

and steel trade from the early postwar period until 1974, culminating in the introduction of the first major postwar steel trade barriers, the voluntary export restraint agreements. Chapter 6 focuses on the pivotal year of 1977, the new protectionist policy devices developed to cope with the new crisis, and the spread of such policies from the United States to the European Community. Chapter 7 examines the most recent steel trade policy developments, which have resulted from the continual renewal of the adjustment crisis fostered by protectionism. A summary of the progressive refinement and spread of these policies, as well as a concluding overview of and outlook on the issue of steel trade policy, is provided in Chapter 8.

This study has benefitted from the generous support of many others. The author received early encouragement in his research from Jan Tumlir and Richard Blackhurst of the General Agreement on Tariffs and Trade, and from Henryk Kierzkowski and Gerard Curzon of the Graduate Institute of International Studies in Geneva. Ingo Walter of New York University, Hans Mueller of Middle Tennessee State University and Gary Horlick, former US deputy assistant secretary for import administration, Department of Commerce, provided many helpful suggestions and comments. The Babson College Board of Research provided generous funding. The author is ultimately indebted most, however, to his wife Tonya and daughter Ana-Lisa, for their constant support and inspiration during the research and writing of this study, which is gratefully dedicated to them.

List of Figures

List of abbreviations

AISI	American Iron and Steel Institute
BOF	Basic Oxygen Furnace
BPS	Basic Price System
CBO	Congressional Budget Office
EC	European Community
ECSC	European Coal and Steel Community
EFTA	European Free Trade Association
ENA	Experimental Negotiating Agreement
FTC	Federal Trade Commission
GATT	General Agreement on Tariffs and Trade
IISI	International Iron and Steel Institute
ITC	International Trade Commission
OJEC	Official Journal of the European Communities
OTA	Office of Technology Assessment
TPM	Trigger Price Mechanism
USW	United Steelworkers
VER	Voluntary Export Restraint
VRAs	Voluntary Restraint Agreements (1969–74)

List of Tables

For Tonya and Ana-Lisa

1

Steel Trade and National Welfare

Steel is at once a myth and a mystique, the symbol of power and prosperity. The nation sees itself in its steel industry, whose interests are confounded with the national interest.

Yves Agnes
'Le Mythe de 1'Acier'
(*Le Monde*, 12 October 1980, p. 1)

LOOK FOR INTRO π

Introduction

For industrialized or industrializing countries of the world, steel is more than an industry; it is a state of mind. The metaphorical content of the word in linguistic usage has direct application in a political context. Steel is economic strength, the symbol of burgeoning industrialization; it is the material of which bridges, cities and manufacturing industries are built. Steel abounds in dramatic industrial imagery: billowing smokestacks, sinewy workers, plants and equipment of Brobdingnagian proportions. It is, furthermore, the key to military defense, a bulwark against an uncertain and hostile world, a provider of tangible security in the form of aircraft, ships, tanks and guns. Steel is, finally, the product of a vital, robust workforce, whose muscle becomes transformed, through the plates and bars and wire it produces, into the national foundation of economic power and growth.

Such is the mystique that surrounds the description of steel as a 'basic industry' and provides a key to understanding its ability to achieve far-reaching influence among policymakers in times of threatened import competition. Its political impact has been reinforced by the development of steel industries as distinctively national entities, by the large economic impact of integrated steel plants on the regions where they are located, and by the sensitivity of the industry to business cycles. Thus, a downturn in the domestic steel industry typi-

cally provides highly visible, dramatic evidence of a downturn in the entire economy: massive layoffs concentrated in specific regions, idle industrial capacity and spillover effects in local communities. The steel industry has, through its stark association with economic conditions in general, become an indicator of national economic well-being.

It is by this process that steel has entered into the national political consciousness and exhibited the singular ability to force the implementation of protective trade restrictions, substituting political for economic channels of adjustment to changing economic conditions. The national character of steel industries in an environment of severe international competition provides the basis for reflex protectionist actions when imports impinge on domestic steel markets. Even if the industry's problems lie largely in domestic market shifts, technological change, or mismanagement, imports provide a highly visible, politically expedient scapegoat on which to shunt the blame. It is in keeping with the mystique of steel that the mundane requirements of internal market and managerial adjustment are often dismissed as largely irrelevant; the contraction of an industry so vital to national prestige, national defense and national economic growth must be traced to influences – indeed, to nefarious forces in many cases – beyond national borders. In times of adversity, economic or otherwise, it is not uncommon to perceive foreign influences as the cause of the problem. Thus, steel imports emerge as the culprit in the decline of the domestic industry. During the crisis year of 1977, it was Japan, the inveterate trade 'disrupter' of recent times, that bore the brunt of these xenophobic tendencies:

> On December 7, 1941, the Japanese bombed Pearl Harbor and killed thousands of Americans. Now it is September, 1977 and the Japanese are still bombing the United States in a different manner. Now they are using steel plates [and] structural steel ... The whole gist of this is that foreign imports are putting steel workers out of business. Foreign imports have invaded every phase of our life. In due time this will probably starve as many people in the United States as the bombs killed on Pearl Harbor.
>
> (Brown, 1977)

In addition to the traces of xenophobia that have appeared in the debate, the steel import issue has also resurrected mercantilist arguments against trade. When it comes to steel, it seems, the logic of international trade theory is suspended. Eugene Frank, steel industry consultant and former US International Trade Commissioner, expressed the view that 'there are no gains from trade' when imports displace local production (Frank, 1984). Trade protection has even

been described as a 'necessary' input, alongside the traditional factors of production, in the manufacture of steel (Yeager, 1980, p. 40). In the emotionally charged atmosphere surrounding the steel trade policy debate, the economic principle of mutual benefits from international trade to *all* participants yields to a seemingly zero-sum game in which the exporting countries win and the importing countries lose. The foreign supplier of steel becomes the enemy, or at least the adversary, as imports become synonymous with layoffs, plant closings, foreign dependence and 'de-industrialization.'

IN ARTICLES AS WELL

Steel protectionism: arguments and legal channels

In spite of the notion that steel is somehow a special case, deserving of special government support and protection from international trade, a critical review of the role of steel in the economy and in overall economic welfare casts doubt on the ability of trade restrictions to improve a country's economic welfare. It is important to remember in this context that, notwithstanding steel's association with national economic strength, governments have traditionally shown a certain ambivalence in their relationship with national steel industries. Market concentration and pricemaking power, for example, have often been serious concerns of public policy in steel producing countries. While restricting trade in steel products may aid the steel industry, it works against the critical national goals of a liberal competition policy and control of the general price level. A policy of open trade, in contrast, has the salutary effect of increasing competition and thereby dispersing economic power in the importing country, reducing the ability of domestic steel producers to collude and fix or unilaterally raise prices. In many countries, a policy conflict also arises in reducing pollution that is linked with steel production. Import restrictions on steel may increase local steel output, but they will also increase the level of pollution. There exists a similar trade-off between protecting a declining steel industry and promoting newer, high-technology industries; by artificially maintaining domestic steel production above its market-driven level, labor and capital are prevented from moving to more efficient, growth-oriented industries.

The role of steel in a country's welfare can be put into perspective in the following manner. Proponents of protectionist policies for the steel industry generally argue that leaving the domestic steel market open to import competition would damage national economic or political interests. Implicit in these arguments is the view that an import-induced contraction of the industry below some presupposed minimum level of output (or employment) would harm national defense capabilities or

national economic health. An ancillary argument rests on the contention that foreign imports are in some sense 'unfairly' traded, particularly with 'predatory' intent, whereby imports threaten to deprive domestic consumers of a local industry, leaving them open to foreign monopoly pricing in the future. From an economic perspective, such arguments claim that there exists a *divergence* between the *private* market cost of steel imports and their *social* cost as determined by either economic distortions or non-economic national goals. A valid argument for a policy of protection from imports on the grounds that it will improve the country's welfare, however defined, requires evidence that (1) a divergence actually exists and (2) the trade restriction will actually correct the divergence without severe over-shooting. In addition, the trade restriction will not be the *best* policy choice if another, less distorting policy (usually a domestic subsidy) provides a practical alternative.

In general, an economic analysis of policy alternatives follows the *rule of specificity*, which requires that government intervention, if it is justified at all, attack a market failure at its root. If the source of the problem lies in the inability of the market to provide a certain 'strategic' level of steel output, for example, then the best policy is one that will increase production to the desired level without causing further distortions in the economy. Such a situation suggests that a direct production subsidy would be the best policy; an import restriction would be an inferior policy because it would introduce a price distortion (a price hike due to the artificial restriction on import competition) in the steel consumer market where none is called for. The source of the problem, in other words, is not imports, *per se*, but an insufficient amount of domestic production. Using this analytical framework, the following discussion will briefly outline the major arguments used to advocate trade restrictions on steel products.

The infant industry argument

Historically many of the roots of steel protectionism grew out of what became known as the infant industry argument, first articulated by Alexander Hamilton (1791) and later developed by the German economist List (1841). According to Hamilton's reasoning, an open trading system would be detrimental to the developing industries of post-colonial America because established foreign firms, with their greater experience and expertise, had already acquired an advantage in both quality and price against which American enterprises could not compete. Furthermore, Hamilton noted that foreign firms often received export subsidies from their governments or formed export cartels, aided and abetted by their governments, with the express purpose of underselling and destroying new producers abroad. Infant industries

· thus deserved temporary protection to overcome their inherent disadvantages and repel such predatory behavior by foreign firms and governments. Once the infant industry became established, however, the protective tariff would no longer be needed. Iron and steel manufactures were singled out in his report articles 'entitled to preeminent rank' because of their importance in the making of farm implements, tools, firearms and construction materials; they were therefore entitled to additional tariff protection in order to keep out imports and further encourage domestic production (Hamilton, 1791, p. 79).

IMPORTANT

From an economic point of view, a valid application of the infant industry argument depends on the existence of a market failure within the national economy, that is, the inability of private markets to efficiently allocate resources. If, for example, a newly formed steel industry would be successful on its own but for an initial lack of experience, skill acquisition, or opportunity to achieve economies of scale, and if these deficiencies cannot be rectified through private markets, then the economy of the country as a whole may be better off by imposing a temporary tariff to correct the market failure, thereby allowing the 'infant' to grow to viable adulthood. The economic test for a legitimate use of an infant industry tariff is a severe one. Normally, for example, firms or individuals could avail themselves of private capital markets to borrow during the 'infancy' period to finance training, initial losses, etc., all in anticipation of future successes that would justify the original investment. If there are imperfections in the capital, labor or information markets, then the best solution theoretically would be to correct the failures in *those* markets through subsidies or other government policies, not through tariffs against imports (Corden, 1974, Chapter 9).

The practice of infant industry protection, furthermore, is rife with difficulties, both theoretical and practical. First of all, a government needs to have some knowledge (presumably absent in private markets) that the industry to be protected will in fact eventually be viable on its own. The government must, in other words, have the ability to pick 'winners.' In addition, there is the political difficulty of removing the protective tariff once the infant 'grows up,' whenever that is. Finally, in order for the country as a whole to be better off as a result of infant industry protection, its benefits must outweigh its costs, a difficult calculation to make before (or even after) the policy is implemented. In other words, a 'successful' infant industry tariff used to protect a presently well-established industry may in fact have left the country worse off than if protection had not been used (Grubel, 1982).

- A variant of the infant industry argument that has emerged in the steel debate is the *senescent industry argument* for protection (see

Gray, 1973). This argument defends trade restrictions as a means for preventing an established industry with obsolete capital from being driven out of business by foreign competitors with modern, cost-effective equipment and technologies. Protection from imports would allow the 'senescent' industry to raise prices, increase profits and thereby finance the modernization program necessary to regain international competitiveness, at which point the trade restrictions could be lifted. Both American and European steelmakers have argued that such price increases are necessary in order for the industry to return to competitive health. In this context, one may legitimately ask why the industry allowed itself to fall into a state of decline. From a public policy perspective, for instance, protectionism in this case may merely be bailing out poor management and planning. Regardless of the reasons why the industry reached this state, however, the same economic objections apply here as under the infant industry argument. If the modernization will truly benefit the steel firm so much, private capital markets should be willing to lend the funds and protection would not be necessary. If capital markets are not functioning properly, then government intervention should focus on improving them, not on distorting trade. If marginal steel firms are unable to raise the necessary modernization funds in a market environment, on the other hand, then from an economic point of view the social value of the investment does not equal the social value of its return, and the modernization should not take place for these firms, even if they must contract or exit the market. Even if there exists a case for government aid to the senescent industry, furthermore, a direct subsidy will be superior to trade restrictions. Crandall (1981, p. 139) has calculated that the cost to the economy and particularly to consumers of import restrictions sufficient to provide funds for comprehensive plant modernization in the United States would be unjustifiably high.

Notwithstanding these considerations of policy efficiency, the thrust of the infant industry argument as applied by most governments has had much more to do with economic nationalism than with economic welfare. This has been particularly apparent in the steel industry, with all its nationalistic trappings. Steel, as was shown earlier, has come to be associated with the national goals of independence, national defense, industrial strength and economic well-being. In a world of national rivalries, suspicions and hostilities, the infant industry argument has taken on the role of a bulwark against the foreign competition that would endanger these goals.

Countercyclical protection

In addition to infant industry considerations, the vulnerability of the steel industry to cyclical downturns has provided an additional argument for

trade protection. Throughout the years, steel markets have proven to be particularly sensitive to business cycles. Demand for steel is based largely on demand for steel products, which have always included mostly capital and other durable goods, such as machinery and construction materials. These are products whose purchase is often delayed when business conditions are bad. During a recession, the demand for steel is therefore likely to drop very sharply. At the same time, steel production involves a large amount of fixed capital cost, and fixed cost per unit will decrease as production increases. If total demand for steel is too low to absorb the industry's output, firms with unused capacity then have the incentive, as long as competition exists in the market, to lower their offer price in order to increase their capacity utilization, lower their fixed cost per unit, and attempt to achieve at least a normal rate of return.[1] However, when all steel firms are pursuing this strategy, the resulting increase in industry output may glut the market, depressing the price well below the level where a normal rate of return is possible for anybody, and perhaps below the 'shut-down' price for some firms.

When a domestic steel industry's steel market is open to international competition, a business downturn abroad typically encourages the foreign producers to use the export market as a means of maintaining output. The resulting downward pressure on steel prices in the import market is unwelcome to domestic producers in any case, and is particularly objectionable if business downturns are synchronized between the trading countries. In this context, a tariff serves the purpose of insulating the domestic market from the surges in imports that may result from international or foreign recessions.

From an economic viewpoint, an anti-recessionary tariff generally benefits domestic producers at the expense of domestic consumers and economic efficiency as a whole. Unless the periodic import surge in itself reduces competition and total economic welfare in the importing country, a countercyclical tariff will not benefit that country economically. Economic efficiency usually requires, in other words, that markets be allowed to clear without government intervention both during boom times, when prices are high, and during downturns, when prices are low.[2] Furthermore, a tariff cannot really eliminate the basic tendency of steel prices to fluctuate over the business cycle, since it does not alter the incentive of producers to maintain capacity utilization during downturns.

Unfair trade practices

Another argument for protection against imports, related to both the infant industry and countercyclical arguments, focuses on acts of dumping and subsidization. The GATT specifically allows govern-

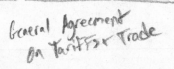

General Agreement
on Tariff for Trade

ments to impose duties on dumped and subsidized imports. The classical definition of dumping is the practice of price discrimination between international and domestic markets or between two or more export markets. An alternative definition of dumping, based on pricing below the cost of production, will be discussed in Chapter 6. Viner (1923), in a comprehensive examination of the subject, noted that dumping may take place for various reasons, such as eliminating overstocks, establishing market shares abroad, or simply maximizing joint profits in two separable markets.[3] There also exists, however, the theoretical possibility that dumping could be used to eliminate foreign competition. According to the predatory scenario, the dumping firm could charge monopoly prices at home and thereby be able to finance a strategy of underselling its foreign competition to the point of driving it out of business. Afterwards, the dumping firm would supposedly be in a position to raise prices to monopoly levels in the 'conquered' market. Yet it is worth noting that successful predation would require not only that the dumping firm have a 'deep pocket' to finance the losses incurred while conducting its price war but also that it somehow suppress competition from other rivals on the world market and from a potential resurgence of the industry in the targeted foreign market.

A similar analysis applies to the argument for countervailing duties against imports subsidized by their home governments. Such allegations were particularly prominent during the American steel industry's 1984 campaign for trade protection. Theoretically, the subsidized import could displace local production, and in extreme cases, eliminate competition, as under the dumping scenario. Yet even if it can be shown both that the alleged subsidies have significantly lowered import prices and that domestic firms have suffered injury as a result, it is unlikely that the subsidy will damage the economic welfare of the importing country as a whole, since the gain to consumers from the lower prices generally outweighs the loss to producers.

None the less, antidumping and countervailing duty laws are based on the idea that demonstrable injury to a domestic industry due to 'unfair' trade provides valid grounds for trade restrictions. Their legitimacy as tools of trade policy derives from their long-standing recognition in national trade laws, reinforced by their inclusion in the GATT. From an economic viewpoint, restrictions on 'unfair' trade can only be justified if their net impact on trade relations is to facilitate trade liberalization in general. By supporting restrictions on dumped and subsidized trade, it is argued, a government interested in trade liberalization will provide a 'safety valve' for protectionist pressure, appeasing otherwise protectionist lobbies and allowing the implementation of liberal trade policies (Lloyd, 1977, pp. 14–15). While laws against 'unfair' trade appear to have served this purpose in the

past, their application to steel trade has more recently indicated their vulnerability to protectionist manipulation, as will be shown in Chapters 6 and 7.

Escape clause protection

Another legal means of restricting trade is the 'escape clause' (article XIX) of the GATT, which allows a country to impose temporary trade protection for an industry that has suffered 'serious injury' due to a surge in imports. The conditions and criteria for a successful application for escape clause protection are based on a legal determination of injury, import penetration and the causal link to imports (see Adams and Dirlam, 1977; Berg, 1982). Originally, the intention of article XIX was to appease potential domestic opposition to trade liberalization, a strategy similar to that of GATT's antidumping provisions, since it indicated that the surge in imports be the result of negotiated tariff reductions. The principle of escape clause protection as represented by national trade laws has come to be applied more broadly, however. Like the unfair trade laws, it also acts as a sort of 'safety valve' for protectionist pressure.

Yet the granting of protection to 'injured' producers necessarily imposes a cost on the economy of the importing country as a whole. Again, the only economic justification for allowing protection under these circumstances would be if it were to make possible an overall liberalization of trade. As originally envisaged by the founders of the GATT, article XIX was intended to allow protection only at a very high price, since it required any escape clause protection to be applied in a non-discrimatory manner and provided for compensation to affected exporters. In an environment of increased protectionist sentiment, however, these stringent conditions have prevented the escape clause from effectively controlling the spread of trade restrictions. Like the unfair trade laws, the legal process of an escape clause investigation has become subject to protectionist manipulation, as was the case in the US steel industry's petition in 1984, to be discussed in Chapter 7.

Non-economic national goals

Finally, many arguments for steel trade restrictions appear to be based on national policy goals, for which it is presumably necessary to restrict imports. The national defense and 'deindustrialization' arguments, as well as regional economic development and income distribution and 'jobs' arguments for protection imply that 'excessive' imports prevent the realization of critical goals. Economic analysis, to be sure, cannot assign values to non-economic goals. However, it can indicate the

potential *cost*, in terms of foregone national economic welfare, of pursuing such goals through protectionist policies.

Import protection for the steel industry based on national defense, for example, can be considered a sort of insurance policy, guaranteeing usable steel capacity if foreign supplies were cut off. Several considerations cast doubt upon the utility of such a policy, however. From a practical viewpoint, trade protection may in fact directly contradict the goal of national security by allowing an uncompetitive industry to avoid adjustment. In addition, the proposal assumes that steel requirements by the military in times of war will exceed the market-determined steelmaking capabilities of the industry in peacetime. Yet in the case of the United States at least, the evidence suggests that potential defense needs do not require the maintenance of capacity in excess of free market levels. Even at the height of the Vietnam War, steel deliveries for military ordnance amounted to just 2.2 per cent of total shipments (AISI, *Annual Statistical Report* 1976, table 16). In addition, the insurance 'premium' of trade restrictions, represented as their economic cost in terms of lost efficiency, may be quite high and may not even be necessary. In an increasingly competitive world steel market, production is dispersed among several countries, including many that are friendly to the United States and western Europe. It is therefore unlikely that steel supplies could be effectively shut off to any particular country in times of a crisis. Finally, a subsidy to the industry will generally represent a less costly alternative if government intervention is in fact necessary (see Crandall, 1981, pp. 98–103).

Similarly, any national goals that establish the need for minimum production levels in the steel industry or specify plant location requirements, can be achieved with less distortion and cost through subsidies to the corresponding economic activity rather than through import restrictions (Corden, 1974, Chapter 2). This applies as well to the most emotional argument for protectionism: that imports 'destroy' jobs in the domestic economy. To be sure, import restrictions on steel can 'save' steel jobs, at least temporarily, but the ability of such measures to rescue workers from a declining steel market has proven to be limited (see Crandall, 1981, p. 139; ITC, 1982b, pp. 69*–70*). At the same time, as noted earlier, by artificially diverting labor and capital towards the steel industry, the country prevents other (usually growth-oriented) industries from developing. Job losses may not stop here, however. By artificially raising the price of steel to domestic steel consumers, steel trade restrictions increase the costs of production for steel-using industries, making them less competitive on world markets. This may cause layoffs in these industries in excess of the number of jobs 'saved' in primary steel production. To these considerations

one must add the possible job losses from foreign retaliation and protection-induced exchange rate distortions in domestic export-oriented industries. In short, import restrictions cannot on balance 'save' jobs, they can only distort the pattern of employment, causing factors of production to remain in uncompetitive sectors and imposing severe economic costs on the economy as a whole.

The political logic of trade restrictions

The case for trade restrictions on steel imports on economic grounds is generally weak, due to the difficult task of identifying a market failure associated with steel *trade* itself. If national policy goals call for a certain level of steel production, then a direct subsidy provides the most efficient means of government intervention. Steel tariffs may be justifiable on legal grounds, however, if a violation of antidumping or countervailing duty laws occurs, or if the conditions for escape clause relief exist. In these cases, a legitimate application of trade laws requires that certain evidence exist, such as a determination that dumping or subsidization, as well as injury in most cases, actually took place.

If steel firms could receive aid in the form of subsidies and a strict application of the trade laws instead of through protectionist policies, why then do they not depend on these channels alone? The answer lies in the political dynamics of government intervention. Trade restrictions are in fact a form of subsidy to the protected industry in that they cause the level of imports to drop, domestic production to increase and the price of the protected product to rise. Unlike direct subsidies, however, trade restrictions are 'off-budget;' they conceal the subsidy in the form of price increases and tend to reduce public exposure to their operation. While direct subsidies would require an embarassing public debate over specific outlays and the appropriateness of allowing the industry to feed at the public trough, trade restrictions provide aid in a more passive, less visible manner. Unlike subsidies, they also have a tendency to persist, removed from public review and scrutiny. The steel industries of the United States and the European Community (EC), which have received nearly continual trade protection for several years, could probably not have acquired the implicit subsidies directly if the issue over renewal had to be debated publicly year after year.

The unfair trade laws, on the other hand, do in fact provide protection from imports, but the decision is based on technical criteria and may only result in small duties. Furthermore, they can generally only provide protection on limited steel product lines for each individual

investigation. In other words, the antidumping and countervailing duty laws, as well as escape clause protection in most cases, cannot provide reliable, comprehensive relief from foreign competition. The industry thus typically appeals for broad trade restrictions attainable through political influence. As the analysis of Chapters 5 through 7 will show, however, trade law channels of protection can act as a useful pawn in a protectionist campaign.

In the end, the calls for protection from the steel industry have thus rarely appealed to strictly economic or legal criteria. They have rather appealed to the idea that increased steel imports, in and of themselves, are 'unfair' or 'undesirable,' in that they represent a foreign threat to a basic, national industry. By moving the debate into the political arena, the industry can apply its formidable political clout to seek influence within policymaking institutions. A successful protectionist campaign will put domestic steel producers in a position to alter otherwise exogenously determined market variables, such as the degree of competition and steel prices. This serves the steel firm's goals of increasing profits and avoiding or delaying the severe economic requirements of adjustment to competitive change, and has provided the industry with a motive for seeking trade restrictions throughout its history.

Notes

[1]   In other words, a steel firm facing decreasing demand will generally be operating under conditions of decreasing average cost. If current revenues are not covering total cost and if demand is sufficiently elastic, the firm may lower its offer price in the expectation of expanding its out put and increasing revenues to the 'break-even' point.

[2]   This argument depends on assumptions regarding market structure and its relationship with the degree of risk aversion. See Sandmo, 1971, Barcelo, 1979, p. 72.

[3]   A firm facing two separable markets, domestic and foreign, will maximize profits by setting two different prices, the lower price applied to the market with higher price elasticity. Since export markets are usually more price elastic than domestic markets, export prices are typically lower than domestic prices when market separation is possible. For a more thorough discussion, see Corden, 1974, Chapter 8.

2

Steel Protectionism in Historical Perspective

The protection of steel producers from import competition has been an issue in international economic relations for as long as steel has been traded. In fact, many of the issues in steel trade relations today — dumping, subsidies, increasing cartelization and technology transfer, to name a few — are really renewed versions of old disputes. Throughout the history of steel trade, periods of rapid technological advancement, overcapacity, or competitive shifts have always resulted in pressures for protection. It is therefore useful to put the current course of steel trade relations in perspective by considering the record of trade restrictions in the past.

The historical pattern of steel trade protectionism reveals two broad cycles. The first dates back to around 1880, when steel trade began in significant quantities, and lasted until the outbreak of the second World War in 1939. The second cycle began in 1945 with the beginning of the postwar ere of multilateral trade relations and continues to this day. Each cycle has been characterized by increasingly sophisticated and restrictive barriers to steel trade. In the mid-1980s, the protectionist cycle was again reaching a peak.

The early years of steel trade

The birthplace of the modern steel industry was Great Britain. Its leadership in steel production and technology was a natural outgrowth of its dominance in ironmaking, which dated back to the mid-eighteenth century. As the leader of the industrial revolution, Britain enjoyed an environment of rapid industrial growth conducive to technological advancement and innovation, giving British steelmakers a decided advantage over their international rivals for many years. British steelmakers in fact, dominated world production until 1886 and led in total exports of iron and steel until 1901.

The early years of modern steel production were also coterminous with the era of free trade in British economic policy. The Corn Laws and Navigation Acts had been abolished in 1846 and 1847, respectively, and by 1860 tariffs on all goods had been abolished (McConnell, 1943, p. 124). For several decades to come, British industrial pre-eminence and free trade policy became inextricably linked in the minds of British policymakers and the public at large. Until the end of the Victorian era, the virtues of free trade were virtually uncontested. One writer of the period noted in fact that '... when I was asked in 1880 ... to write something in defence of Free Trade, it seemed to me ... as if I had been asked to prove Euclid, or give a reason for the rules of Grammar' (T. H. Farrer, 1886, quoted in Carr and Taplin, 1962, p. 122). It is clear that the dominance of British steelmakers in world production and trade provided an auspicious environment for free trade thinking. With no serious international rivals for many years, the free importation of steel products posed no threat to the industry.

The British adherence to a policy of free trade stands in sharp contrast to the protectionist trade policies of the other major steel producers of the period, the United States and Germany. In the United States the steel industry was heavily protected by tariffs until the end of the nineteenth century. Germany, after a brief period of free trade in the late 1870s, imposed tariffs of 25−30 Deutschmark per ton on most steel products in 1879, which remained basically unchanged for the next 60 years. The *ad valorem* equivalent of these tariffs ranged from about 10−20 per cent for the various protected steel products, based on 1879 prices (Kestner, 1902, pp. 50−53). Other countries also protected their steel industries from imports, although the degree of protection varied. Belgium, whose steel industry was small but efficient and competitive on export markets, imposed a low tariff, while France, whose steel was uncompetitive on world markets at this point, imposed high tariffs on imports.

The widespread use of trade restrictions on steel among the industrializing countries of this period was based on three inter-related arguments. Initially, the infant industry argument for protection was most important. For the young steel industries in the United States and Germany, the well-established British industry appeared capable in the 1870s and 1880s of crushing any potential competition by flooding its domestic market with cheap steel. A second argument for protection grew out of the vulnerability of steel producers to cyclical downswings, which severely depressed prices and led to surges in imports. A third argument, related to the first two, relied on the allegedly damaging impact of dumping by foreign steel suppliers on domestic producers. Each argument, in essence, called for tariffs to counteract the predatory trade practices of foreign steel producers.

Fear of predation and the drive for industrial independence
The fear of predatory trade practices was particularly strong in the United States, whose steel industry grew up in the shadow of its British counterpart. At a time when the United States was striving to attain total economic independence from Great Britain, the predatory scenarios associated with the infant industry, countercyclical and anti-dumping arguments remained very influential in American trade policy in general. As an instrument of economic nationalism, its popularity was due in large part to its consonance with assumptions made about the predatory motives of British exporters (Viner, 1923, p. 42). Such fears were reinforced by statements made in Great Britain regarding the sacrifices of her exporters. The British commentator Henry Brougham, for example, asserted in 1816 that

> it was well worth while to incur a loss upon the first exportation, in order, by the glut, to stifle in the cradle those rising manufactures in the United States which the war [of 1812] had forced into existence contrary to the natural course of things.
> (*Edinburgh Review*, 1816, pp. 263–4, quoted in Viner, 1923, p. 42)

While Viner (1923, p. 42) has noted that Brougham's statement was based more on wishful thinking than on fact, one author observed that 'Mr. Brougham's words have often since done their duty in firing the protectionist heart' (Stanwood, 1904, vol. I, p. 168).

Furthermore, British iron exports appeared to be geared specifically towards the domination of foreign markets, as described in an official report from 1854, which indicated the 'immense losses' that iron producers suffered in bad times 'in order to destroy foreign competition' (*Report to the Commissioner ...*, 1854, p. 20). While this statement has been interpreted as an open admission of predatory export practices (see Hogan, 1971, p. 173), a further reading suggests that it was part of an argument to persuade British labor to refrain from damaging strikes, which, if they took place, would mean that 'the great accumulations of capital could no longer be made which enable a few of the most wealthy capitalists to overwhelm all foreign competition in times of great depression' (Hogan, 1971).

Despite the lack of hard evidence that British exports were actually capable of suppressing the American iron industry, heavy duties were applied to competing imports in the first half of the nineteenth century (Taussig, 1931, p. 52). The United States thus entered the steel age with a heavy legacy of protectionism. This tradition would be reinforced by the fact that its perennial rival, Great Britain, once again had the advantage in technology, experience and productive capacity in this new product.

In Germany, protectionism was not so firmly entrenched as steel trade began, and in the 1870s tariffs were in fact reduced as the free trade philosophy of Great Britain made its way into the new German empire. By 1877 iron and steel products were admitted into Germany duty free. However, technological and political factors combined to reverse Germany's open trading policy. The late 1870s witnessed the development in Great Britain of the Thomas converter process, which greatly facilitated the use of phosphorus-rich iron ore. The new process was of particular interest to German steelmakers, since most of Germany's iron ore supplies had a high phosphorus content. A variant of the infant industry argument, which called for a 'learning tariff' (*Erziehungszoll*) seemed applicable here: in order to adopt and exploit the new technology before British steelmakers could damage the German industry, some temporary protection was necessary. Other (mostly political) factors reinforced the case for protection. The German steel industry wielded considerable influence in the *Reichstag* and seemed to be as concerned in protecting its outdated blast furnaces as it was in adopting the Thomas converter. For Bismarck, the national defense argument was probably more compelling than any economic arguments: he viewed the protection of the steel industry as a means of building a strong military based on secure sources of domestic supply (Goldstein, 1912, pp. 18–22). In 1879 Germany reintroduced tariffs on iron and steel products.

Did the steel tariffs work?
It is useful in this context to consider the efficacy of tariffs in promoting steel production in what would become the two steel giants, the United States and Germany. Certainly, their steel industries grew impressively during the period of protection, just as the infant industry argument predicted. Yet there is much evidence that they would have grown even without protection. Steelmakers in both countries assimilated and improved upon new technologies from Great Britain and exploited the presence of the important ingredients for successful steelmaking: access to iron ore, coking coal, capital, manpower and markets to absorb production. It is likely that the most the tariff did was to speed up the development of the steel industry in both countries (Taussig, 1915, Chapter 10; Berglund and Wright, 1929, pp. 130–5). In this regard, the extended lifespan of infant industry protection also points to its welfare costs to the economy. Tariffs helped to reduce competition on both the German and American domestic markets, making possible unusually large profits. In the case of American steel rails, for example, returns on investment were estimated to be 100 per cent or more (Taussig, 1931, p. 222). The reduced consumption and higher prices caused by the tariff created an economic welfare cost that

must be subtracted from any benefits resulting from the tariff in a calculation of its net impact on the economy. The longer the tariff was in effect, therefore, the larger its welfare cost was to the country.

Tariffs, dumping and trade relations
The rapid development of the steel industries of Germany and the United States soon diminished the validity of infant industry protection in these countries. Yet a predictable problem with the use of such protectionist policies was the difficulty of removing them once their original purpose had been eliminated. The political battle to maintain protection for the steel industry thus turned to the antidumping argument as the basis for continued tariffs.

As was mentioned earlier, one of the major roots of protectionist sentiment in the nineteenth century was the fear of predatory dumping. The irony of the use of tariffs to protect against dumping, however, was that the tariffs also made it possible for domestic steel producers in Germany and the United States to practice dumping of their own. In order to successfully dump a good in a foreign market, it is necessary to isolate that market from the home market; otherwise a process of price arbitrage could take place, in which the cheaper dumped goods could be profitably re-imported into the higher-priced domestic market. One such means of separating the markets is the cost of transporting the good from the home to the foreign market, which in the case of steel was a major reason dumping could take place. However, the tariffs provided an even more propitious environment for dumping, since they gave domestic steel producers an extra margin by which to practise price discrimination. This link between tariffs and dumping thus gave domestic steel producers an added incentive to seek continued protection from imports.

Both American and German steel exporters practised dumping from behind their tariff walls in the late nineteenth and early twentieth centuries (Viner, 1923, Chapters 4 and 5). There is no evidence that the dumping actually succeeded in eliminating competition, but with a few exceptions, it appears that the motive was not predatory but rather simple profit maximization in foreign and domestic markets (Viner, 1923, p. 64). The main damage done by dumping, however, did not result from its effect of lowering import prices, but from its power to spread demands for protection to other countries.

The singular ability of dumping to poison trade relations was revealed in the early years of steel trade. The nearly universal condemnation of dumping has always been perplexing to economists because the importing country almost surely stands to gain from it, a fact recognized by commentators of the time (Gothein, 1904). However, the mere possiblity of a predatory motive, regardless of its chances of

success, has always seemed to have the effect of removing economic criteria from the discussion. In viewing international economic relations from a nationalistic perspective, governments tend to regard home markets as national property rights, whereby access from abroad is automatically regulated and imports are subject to 'fairness' criteria usually defined by the interests of domestic producers. This view emphasizes the dangers involved in the foreign penetration of domestic markets and seeks to assure that pricing follows rules favorable to domestic producers. Low prices that do not conform to the rules are thus viewed as an attack, not only on the import market, but on the importing country's economic welfare and independence as well. This deep-seated fear of economic domination by foreigners is symbolized by the antidumping duty.

In this manner, dumping by German and American steel producers provided both countries with a convenient justification for continued steel tariffs based on the predatory scenario. At the same time, the tariffs themselves created an environment for their own steel producers to dump. The combination of tariffs and dumping practices in these two countries also threatened to spread protectionism to Great Britain in the 1890s and early 1900s. Tariffs had cut off Britain's export-oriented steel industry from these lucrative foreign markets, where British steel could otherwise have competed, while tariff-supported dumping was contributing to increased import penetration in the British domestic steel market from these two sources beginning in the 1890s, amplifying the depression of prices during cyclical downswings.

The greatest challenge to Britain's free trade policy during this period occurred during the extended depression in the domestic steel industry from 1900 to 1904. Cheap and allegedly dumped imports were seen by many to be the cause of the low prices and massive closures and layoffs. An unofficial 'tariff commission' was established by members of the opposition Unionist party to investigate the need for tariff protection in the iron and steel industry. The details of the commission's report appear amazingly familiar to observers of current steel trade relations: reports of layoffs and poor morale among remaining workers, widespread sentiment against 'unfair' trading practices of foreign steelmakers, illustrations of alleged dumping by foreigners at below production cost, and fears that foreign producers would destroy the domestic industry and then raise prices to monopoly levels. The commission concluded by recommending the establishment of tariffs on iron and steel products ranging from 5 to 10 per cent (Carr and Taplin, 1962, pp. 197–203).

In the end, however, the protectionist campaign failed. After the 'commission' findings were published, economic conditions improved dramatically, and the elections of 1906 re-affirmed the policy of free

trade. The upswing in steel demand seemed to vindicate the proponents of free trade, who had argued that the industry's problems had been caused by cyclical forces and by the need for structural adjustment to diminishing ore supplies, new technologies and expanding world competition, not by cheap imports as such (Burn, 1961, Chapter 12). Yet the rejection of tariffs also indicated the durability of Great Britain's free trade tradition. Whereas in the United States and Germany the fear of predation and the interests of steel producers had predominated, in Great Britain there was a greater fear of the ill effects of a turn to tariff protection: foreign tariff retaliation leading to further tariff escalation, anti-competitive domestic market effects and delayed adjustment. Even the British steelmakers themselves, angry as they were at the trade practices of their American and German rivals, had prospered in earlier times under free trade and were hesitant to support a return to tariffs (Carr and Taplin, 1962, p. 198). In this particular policy debate, pressures for protection thus proved unable to reach a requisite threshold level of influence to dislodge a well-entrenched policy of free trade.

American trade liberalization in the early 1900s

While Great Britain's policy of free trade was being challenged by its steel industry's competitive decline on world markets, the emergence of the United States steel industry as the world leader was moving its trade policy towards freer trade. To be sure, tariffs on most steel imports to the United States remained high until the Tariff Act of 1909 as the steel industry's interests continued to hold sway over trade policy. Yet a call for trade liberalization had already been heard during the debate over the Tariff Act of 1883 and by the early 1900s tariff reductions became part of the platform of both political parties. By 1913, tariffs on most steel imports were either eliminated entirely or reduced to levels of 15 per cent or less (Berglund and Wright, 1929, p. 110).

The unilateral reduction of tariffs by the United States during this time can be traced to two trends: the active and vocal increase in opposition to existing tariff legislation and the decrease in the steel industry's motivation to seek further trade protection. Opposition to the tariffs focused initially on the large profits the steel industry was enjoying behind tariff walls. Later, this concern led to the allegation that tariffs had promoted cartels and other collusive arrangements within the steel industry, although Taussig (1915, Chapter 12) rejected the notion that tariffs were responsible for the formation of cartels in the United States. None the less, anti-trust sentiment of the period, which led to the Sherman Anti-trust Act of 1890, lent further momentum to the tariff reform movement.

For American steel producers, on the other hand, tariffs had become less and less important to the well-being of the industry during this period. The American industry led the world in steel production and had matched or surpassed the cost-efficiency standards of its international rivals by the turn of the century (Taussig, 1915, Chapter 9). Despite reduced tariffs, imports had declined to a trickle and no longer posed a threat to the booming American steel industry. The biggest threat to the industry, in fact, probably lay in the new anti-trust laws, and the 'trust-busting' movement, as noted above, had consciously linked tariffs with monopolization. The American steel industry therefore assumed a low profile on the tariff issue in the early 1900s.

Thus, despite the apparently deeply ingrained protectionist sentiments displayed as the steel age began, trade liberalization in the United States was indeed possible. The necessary ingredients at the time included a strong political movement towards reducing tariffs and a prosperous domestic industry, confident of its ability to compete on world markets. Ironically, these very facts meant that an opening of the American market would have little effect on world steel trade, and without a system of multilateral trade relations, trade liberalization could not spread to other countries. As the world economy entered the cataclysmic years of world war and depression, steel trade policy in all countries became increasingly vulnerable to protectionism.

Increasing instability and protectionism after the First World War

The First World War and its economic aftermath played a large role in the increased use of protectionism in the 1920s. The severe disruption of trade patterns and trade relations during the war itself was only a temporary, superficial effect of the hostilities. The deeper impact of the war lay in the increasingly defensive and protective attitudes of governments towards their domestic steel industries. There was, in addition, a growing disenchantment with the ability of an open trading system to allow stability in an increasingly complex world steel market.

The heightened concern of governments for the health of their national steel industries was a direct result of the importance of steel in the war effort. In general, governmental intervention in the steel industry took the form of production targeting, subsidies, planned allocation of raw materials and price controls (Carr and Taplin, 1962, Chapter 29). Much of this regulatory machinery remained in place at the end of the war. Of more importance, however, was the increased suspicion with which governments viewed steel trade. Trade practices by foreign countries that could potentially weaken a national steel industry's ability to support a renewed war effort were viewed as a real

threat to national security. Such suspicions were part of the general surge in nationalism that followed the war. The past record of dumping by two of the belligerents, the United States and Germany, reinforced these attitudes.

The increase in the number of steel producing countries and the growth in world steelmaking capacity exacerbated fears of disruptive trade practices. As was noted earlier, a cyclical downturn in domestic steel demand tends to encourage an increase in exports, depressing prices in export markets. The potential sources of such disruption were steadily increasing in number as new national steel industries grew. This led to a general pessimism regarding the future of world steel prices at the time, which was in fact borne out by their downward trend in the first years of the postwar period (Burn, 1961, p. 407).

Scenarios of predatory and otherwise disruptive trade practices thus created fertile ground for protectionism at this time. Most countries increased tariff protection for steel from their prewar levels. Even the United States, which by that time had become far and away the largest and most advanced steel producer in the world, reversed its policy of trade liberalization and increased most tariffs across the board, following the worldwide trend towards economic nationalism and greater self-sufficiency (Taussig, 1931, pp. 449–50). The problem was compounded by the collapse of the gold standard during the First World War, causing some countries to use import restrictions to alleviate balance-of-payments deficits. This problem had also led to the increased use of import quotas, import licensing systems and bilateral trade agreements as methods of trade restriction (US Tariff Commission, 1938, pp. 444–5).

The first international steel cartel

During the years of economic uncertainty immediately following the First World War, steel producers became increasingly disenchanted with the ability of existing protectionist policies to alleviate the perennial problems of steel markets: cyclical over-capacity and price instability. The tendency of firms to try to maintain high capacity utilization rates and seek additional export markets during market downturns could not, in this view, be reduced by unco-ordinated sets of tariffs and other trade-restricting policies by governments. The solution, it was argued, lay in organizing the industry into international cartels that would co-ordinate production and marketing arrangements. An agreement on overall production levels in the world steel industry, for example, would theoretically eliminate the problem of price instability as output could be matched with existing market conditions. Marketing agreements would establish and guarantee 'fair' market shares for the members of the cartel.

Economic objections to cartels focus on their artificial restriction on supply, which, as with monopolies, misallocates the economy's resources and reduces the country's total output (GNP). In this regard, cartels are likely to be worse than monopolies because cartels generally require that production cutbacks are to be shared according to some negotiated scheme among the various private firms involved. The resulting economic cost is made worse by the fact that the more efficient firms do not usually get the largest production quotas, as would be dictated by cost-minimizing criteria. Instead, politics tends to supersede economics in deciding the pattern of production quotas. By subjecting the determination of market or production shares to a process of negotiation, bickering and arm-twisting rather than impersonal market forces, however, the 'co-operative' aspect of cartels becomes very tenuous indeed. Under such collusive agreements there is always an incentive to secretly cheat on one's quota allocation. If steel prices are propped up by an artificial restriction on total output, for example, it would benefit an individual firm to exceed its quota and sell as much as it could, as long as everyone else remained honest. For firms feeling unfairly slighted in the cartel's quota allocations, the temptation to cheat becomes particularly strong.

International cartels had been previously attempted in particular steel products, but in 1926 the first all-embracing arrangement, the International Steel Cartel (ISC) came into existence. Founding members included Germany, France, Belgium, Luxembourg and the Saar (later to be re-united with Germany). Austria, Hungary and Czechoslovakia joined in 1927. The ISC in its original form had an extremely simple format: fixed production quotas were assigned to each member country and a system of fines was established to prevent overproduction and cheating. This arrangement required that each member country have an effective national cartel that could set production quotas for individual firms. It thus also required at least tacit government approval of what was essentially a collusive agreement among domestic producers.

The reaction of the governments of cartel members to this new form of protectionism was in fact generally favorable, and some, such as German foreign minister Gustav Stresemann, even regarded the ISC as 'a landmark of international economic policy.' Open steel markets, in his view, had only produced self-destructive price wars and predatory competition. 'An atmosphere full of tension caused by opposing economic interests not only conceals dangers for the development of industries themselves but also implies threats to the political peace of the nations. I hail this attempt to dispel these heavy strains . . .' (*Frankfurter Zeitung*, 2 October 1926, p. 1. Quoted in Hexner, 1943, p. 221). Acceptance of the ISC thus reflected the increasing disillusion-

ment among governments with the ability of market mechanisms to serve the economic interests of their countries. Their willingness to rely instead on the decisions of an international bureaucracy of steel producers represented a major turn from economics towards politics as the dominant force in world steel trade.

Notwithstanding Herr Stresemann's high hopes for the ISC agreement of 1926, it ended an utter failure after just 3 years. Its inability to influence steel prices lay in its own structural weaknesses and inflexibility. The agreement had attempted to control the entire steel market with gross production quotas when real control could only be obtained through marketing arrangements in individual finished steel products. The quotas and fines were, furthermore, a source of bickering, especially over the German production share. When Germany withdrew from the agreement in 1929, the ISC was finished.

Depression, protection and the second ISC agreement
The Wall Street crash followed shortly after the collapse of the first ISC agreement, ushering in a decade of worldwide economic depression. World demand for steel shrank drastically, leading to a sharp drop in steel prices that reached its nadir in 1932. The chronic instability of steel prices had returned with a vengence, destroying most of what confidence had remained among policymakers in an open trading system. The threshold level of political influence required to implement or accept new protectionist policies was thereby lowered further. As domestic demand collapsed in the market economies, protectionist fever gripped policymakers, who attempted to generate demand at home and abroad for domestic output through increased trade restrictions and competitive currency devaluations. Such 'beggar-thy-neighbor' policies only succeeded, however, in causing other countries to retaliate with similar policies. Even Great Britain, which had pursued a policy of free trade for 90 years, imposed an import duty on steel of 10 per cent (later raised for most products to 33 per cent) in 1932.

The renewed crisis in steel prices, accompanied by a tightening of export markets through protectionist measures, galvanized the charter members of the original ISC — the steel cartels of Belgium, France, Germany and Luxembourg — into negotiating a new and 'improved' ISC agreement, which took effect in June 1933 and lasted until the outbreak of the Second World War in 1939. The new ISC set out to rectify the mistakes of the old. Domestic production, for example, was no longer regulated under the new accord, which set quotas on exports only. More importantly, marketing agreements for individual steel products were established in order to exercise comprehensive control over steel trade.

The renewed ISC's impact on world steel trade depended on its

ability to include all the major steel exporting firms in the agreement. Most of the major steel industries not party to the ISC in 1933 did in fact eventually participate, including those of Poland, Czechoslovakia, Great Britain and the United States. It is noteworthy that American firms, normally prohibited from engaging in any collusive activities, were permitted by the Webb—Pomerene Act of 1918 to form export cartels, thus allowing their participation in the ISC. On paper, at least, the control of the ISC over the world steel export market and export prices was indeed impressive. By 1938, ISC agreements covered about five-sixths of world steel trade. It restricted steel trade among member countries more effectively than existing tariffs and quotas, since participating producers had pledged to refrain from exporting into each others' domestic markets without prior consent (Hexner, 1943, p. 101). ISC price quotations appeared to hold up and provide some stability in the various steel markets, although the readjustments were often made in response to changing market conditions (Jones, 1979, p. 147).

Yet the ISC was surely much weaker than official statistics showed, and it was often forced to compromise its policies in order to prevent the agreement from breaking apart. The price listings, for example, represented official quotations only. Unauthorized discounts and rebates were evidently common, and were tacitly accepted by other ISC members, who apparently considered such practices preferable to open conflict within the cartel (*The Statist*, 2 July 1938). The 'troublemakers' in the ISC were primarily the cost-efficient Belgian firms, which were extremely competitive and often unsatisfied with their export quotas. Since they were in a position to disrupt the cartel's policies, their influence on ISC quotas and prices was disproportionately large (Jones, 1979, p. 147).

The second ISC agreement therefore proved to be at best moderately successful in propping up steel export prices. While it attempted to gain disciplined cartel control over the world steel export market, it managed to stick together only by moderation in its enforcement of the agreement and by compromise with recalcitrant members. In contrast, the member national cartels appear to have received substantial trade protection from the ISC structures against market interpenetration. Kiersch (1954) has shown that domestic steel prices in Britain, France and Germany lay well above export prices during most of the second ISC agreement.

The ISC thus made little, if any, contribution to international peace and co-operation, as envisaged by many of its supporters. While it was perhaps naive to consider any price-fixing agreement as a contribution to international peace, it was particularly quixotic to expect a cartel in a strategic industry like steel to reduce tensions and suspicions. In

general, cartel politics are based on power, which is concentrated in the larger and stronger participants. In the end, the ISC's contribution to international 'co-operation' in the political sphere was limited to the upward adjustment of German quotas after the Austrian *Anschluss* and the dismemberment of Czechoslovakia. Since bickering over quotas and prices and clandestine discounts appeared to be common, the ISC seemed in fact to contribute to the petrification of existing suspicions, jealousies and distrust that culminated in war.

The postwar period

At the end of the Second World War, the steel industries of most countries outside the United States were in ruins, and trade relations began anew with a clean slate. The postwar period thus began as a time of rebuilding, not only for the steel capacity worldwide, but for the international economic environment as well. The victorious allies, led by the United States, set out to consolidate the tragic economic lessons of the 1930s and construct a new, open system of international trade and payments. Their efforts culminated in the founding of the International Monetary Fund and the General Agreement on Tariffs and Trade (GATT), institutions directed towards increasing and facilitating world trade. The GATT, in particular, sought to reduce trade barriers on a multilateral, non-discriminatory basis. A series of GATT-sponsored trade negotiations reduced steel tariffs of all the major steel-trading countries. By the end of the Kennedy Round of trade negotiations in 1968, steel tariff levels of all GATT member countries had been harmonized at about 7 per cent *ad valorem*, thereby eliminating the general tariff as a means of restricting steel trade (Preeg, 1970, p. 212).

The biggest single ally of the trend towards the liberalization of steel trade was the unprecedented period of economic growth after the war, which was in part a result of the trade liberalization itself (Blackhurst *et al.*, 1977). As long as domestic economic recovery and growth generated steel demand for national steel industries, there was little conflict over steel trade. Under these circumstances, imports were actually welcomed to eliminate domestic steel shortages during cyclical boom periods. The absence of perceived threat in the importation of steel, particularly in the major steel-consuming areas such as the United States and Europe, was an important element in sustaining steel trade liberalization during this time.

As world steel capacity expanded beyond the domestic needs of national steel industries, however, steel imports again became an issue. By the 1960s, the steel industry of Europe had recovered from

the war and the steel industry of Japan had become a major exporter. Both were using the open and lucrative United States market as a major outlet. The world's leading steel producer, the United States, was experiencing a competitive decline, similar to that of Great Britain's steel industry in the late nineteenth century. Yet, unlike Great Britain at that time, the United States possessed no firmly entrenched commitment to an open trading system. The increased penetration of American steel markets by imports represented a threat to the market shares of domestic producers, generating significant protectionist sentiment for the first time in the postwar period. The recession of 1967−8 further depressed prices, lowering the threshold of influence required to enact protectionist policies. The postwar history of steel protectionism thus began with the Voluntary Restraint Agreements, which set quantitative limits on steel exports from Japan and Europe to the United States from 1969 to 1974. European steel producers, fearing a diversion of Japanese exports to their markets, concluded a similar agreement with Japan in 1971.

This period also marked the increased involvement of governments in steel production, particularly in Europe. As prosperity continued, governments turned their attention more and more towards national industrial planning, regional development and social welfare, policy goals for which comprehensive control of the domestic steel industry was to be an important tool. Great Britain, France and Italy nationalized their steel industries, effectively subsidizing investment and even output in some cases. This led other European governments to subsidize their privately owned steel industries in an effort to keep them competitive (Walter, 1979, p. 171). At the same time a number of newly industrializing countries (NICs) were developing their own steel industries under national control. The NICs protected their developing steel industries from imports, using the same infant industry arguments applied by American and European governments a century earlier.

When the short-lived resurgence in steel demand came in the early 1970s, governments as well as private firms were eager to finance the construction of yet more steelmaking capacity, particularly in Europe. The world steel industry was therefore exposed, as it had been in the 1920s under different circumstances, to the problems of overcapacity that accompany a downturn in steel demand. This problem was compounded in the established steel industries of the United States and Europe by a secular decline in domestic steel demand and a decline in the competitiveness of their steel producers on world markets. The steel market downturn sparked by the oil crisis of the mid-1970s was therefore particularly severe, setting off protectionist responses similar in scope to those in the 1920s and 1930s. The European Community

(EC) and the United States imposed a series of escalating trade restrictions, beginning in 1975, that included more voluntary export restraints (VERs), antidumping and countervailing duty investigations and antidumping 'trigger prices.' The EC restrictions concentrated at first on the Japanese but then sought to impose comprehensive controls on all steel deliveries to its markets. The American trade barriers were aimed at first against the Japanese, then against the EC and finally against the NICs in a cumulative effort to impose global import restrictions on the American market. The proliferation of protectionism during the postwar crisis in steel trade has indeed been reminiscent of the retaliatory measures taken in response to 'beggar-thy-neighbor' policies of the 1930s even if the tools of trade restriction have changed.

Another flashback to the steel trade crisis of the 1930s can be seen in the creation of a new international steel cartel, Eurofer, which has had many of the same functions − and problems − as its predecessor, the ISC. The main difference between the two is that Eurofer was founded (and continues to operate) under the direct control of a supranational governmental body, the Commission of the European Communities, while the ISC was in private hands. The official cartel Eurofer was the result of an escalation in EC attempts to rescue the steel industry from the devastating downturn in the market and in its own competitiveness. Earlier stopgap guideline measures and recommendations had proven ineffective; comprehensive, mandatory controls augmented by severe restrictions on steel imports represented the culmination of a process of continually frustrated crisis management. At this writing, the frustration continues.

Summary: steel and chronic protectionism

The preceding historical review has shown that the practice of free trade and trade liberalization have been more the exception than the rule since the large-scale steel trade began. Among the major steel producing countries, Great Britain practised free trade until 1932, while the United States reduced steel tariffs during the late 1800s and early 1900s. Belgium, with its smaller but competitive export-oriented steel industry, has consistently had low tariffs on steel. The United States led the way in multilateral reductions in steel and other tariffs after the Second World War, but protectionism re-appeared as soon as steel trade began to surge in the 1960s. Conditions for open steel trade policies appear to include both a very competitive, self-confident domestic steel industry and a compelling and widespread political appreciation of the merits of liberalized trade, which has generally been tied to the export-orientation of the industry, anti-trust consid-

erations, or a cataclysmic event such as war. In analytic terms, the level of import threat and accompanying protectionist sentiment must be low and the threshold level of political influence needed to impose trade restrictions must be high.

Yet steel industries seem to be particularly susceptible to market instabilities and trade-related disturbances that quickly generate protectionist sentiment. Throughout the history of steel trade, cyclical downswings and the problems of over-capacity, technological lags, shifting cost factors and competition from new international rivals have presented national steel industries with often painful adjustment problems. As an industry perceived to be of such grave strategic importance to national defense and economic welfare, steel has inspired numerous protectionist arguments that would spare the industry from the pain of adjustment. Most of these arguments have focused on an anticipated scenario of foreign predation designed to weaken the importing country's industrial capability, national defense, or employment level.

Within each broad historical trade policy cycle, trade restrictions have furthermore moved progressively from simple tariffs to more rigid, complicated devices such as cartel arrangements. This trend appears to be designed to give domestic steel producers increasing rates of protection and implies that there exists a sort of protectionist 'learning curve,' whereby experience from earlier, ineffective measures is used to design or demand more comprehensive, restrictive trade controls. At the same time, the escalation of protectionist policies in one country has typically involved a proliferation of protectionism among other steel-producing countries. The vicious circle of trade restrictions and retaliation has created two major episodes of spiraling protectionism in steel trade, one in the 1930s, the other beginning in the mid-1970s and continuing to the present day.

The historical legacy of steel protectionism therefore presents policymakers in the major steel-trading countries with a dilemma. To follow the current trend in protectionist policies in steel implies a virtual institutionalization of trade disputes among trading partners, since further steps to create market-sharing arrangements would be subject to the dissension and conflict that accompanies international cartels. Any attempts to reduce current steel trade barriers, on the other hand, would seemingly require heroic efforts on the part of trade officials to overcome the absence of the conditions for trade liberalization mentioned above. The following chapters will explore the reasons for the decline of political will needed to back an open trading system for steel and the prospects for a solution to the policymakers' dilemma.

3

The Battle Lines of Steel
Protectionism

Introduction

The foregoing examination of the history of steel trade policy has
shown that proponents of steel protectionism have consistently and
systematically penetrated policymaking structures to limit competition
from imports. Observation suggests furthermore a general sequence of
events and influences that leads to the phenomenon of steel trade
protection: first, the existence of imports regarded by certain groups as
disruptive or otherwise damaging to their interests; second, the application of influence by these groups to persuade or compel policymakers to grant the desired trade protection; and third, the actual
adoption and implementation of the trade-restricting policy, followed
by associated positive effects, income redistribution and possible foreign retaliation.

A systematic approach to the study of steel trade policy requires an
examination of the process and structures involved in producing protectionist policies. The model of policy formation to be presented in
this chapter comprises three sequential stages or phases of this process,
corresponding to the levels of motivation, generation and effects of
steel protectionism. Stage I begins with a trade-related disturbance
that forces domestic steel firms to adjust in order to maintain a competitive position on world markets. Non-adjusting firms may be motivated to demand protection, bringing the protectionist process to stage
II: the policy debate. Pro- and anti-protectionist influences then come
to bear on policymaking bodies. If a protectionist policy is adopted,
the process moves to stage III; the aftermath of protectionism. Positive
economic effects, as well as international political and economic after-
shocks, follow the policy's implementation. The anticipation of these
effects shapes, in turn, the types of protectionist policies used.

The competitive shock from abroad

The sources of trade disturbances
The formation of policies that protect domestic industries from foreign trade competition can generally be described as the culmination of a *political* process triggered by the failure of firms to adapt to an *economic* process of adjustment. Steel industries have proven to be particularly vulnerable to trade-related disturbances, resistant to adjustment and capable of asserting the requisite level of political influence to avoid market-driven adjustment.

The economic process associated with increased import competition begins when foreign penetration of the domestic market reduces domestic steel profits to a 'critical' level as perceived by the firm or industry. Reasons for the increased import penetration reside in shifts in steel consumer preferences, as well as input cost, technological change and other supply factors.

Price differentials are likely to be an important element in the trade disturbance and usually form the focus of the policy debate. A lowering of the price of steel imports can be caused by short-term or long-term factors. Short-term factors include business cycle fluctuations and other market conditions producing temporary excess capacity. Government policies in the exporting country may also encourage a surge in steel exports, such as restrictions on layoffs, export targeting, or outright subsidies. If such pricing practices involve dumping or other 'unfair' trading practices, the protectionist debate will focus on legally prescribed remedies.

In contrast, long-term declines in import prices point to a corresponding secular decline in the domestic steel industry's competitiveness. This sort of market disturbance is typically caused by either a lowering of production costs abroad relative to domestic costs or an independent increase in export capacity on the world market. Production cost reductions are usually a result of either technological innovations or downward shifts in factor or operating costs. Increases in export supply can generally be traced to either the expansion of existing firms, new entry of firms or long-term shrinkages in the domestic steel demand of exporting countries, a topic to be pursued in Chapter 4. A decrease in the domestic demand for steel in the importing country will also cause the import price to decrease. It is likely, however, that several of these factors will arise at the same time, since they are often interrelated. For example, a technological advance initiated abroad, such as continuous casting or the basic oxygen furnace, lowers production costs for existing firms that adopt the innovation, but may also offer new opportunities for production in low-wage countries. Similarly, the expansion of existing firms may yield econo-

mies of scale, lowering costs. In any case, these disturbances affect the import-competing steel industry by shifting the import supply curve outward.

Domestic market considerations are also important in determining the impact of foreign trade disturbances. The cyclical nature of the steel industry will generally allow the home market to absorb competitive shocks from abroad without much difficulty during a high-demand period, since domestic steel output can be maintained (even if imports are rising), prices remain high and producers are therefore happy. A cyclical downswing, on the other hand, tends to magnify the impact of increased foreign competition, especially when business cycles are synchronized in such a way that low domestic demand in exporting countries shifts their export supply schedules outward at a time when domestic steel demand is also depressed. General price-cutting on top of shifts in cost advantage thus represents the most severe disturbance facing import-competing steel firms.

Price differentials provide only a partial explanation of import penetration, however. For steel consumers choosing among various foreign and domestic sources of supply, a trade-off exists, for example, between the price of steel and the risk associated with reliability of delivery. In general, the longer the supply lines and uncertainty of timely delivery, the riskier is the supply source. Normally, foreign suppliers, especially recent and unproven market entrants, are considered riskier for these reasons and may therefore have to discount their products even more in order to enter the foreign market. Under these circumstances, price differentials favoring the foreign steel supplier may not reflect a competitive advantage, but rather a situation of parity with domestic suppliers. Yet the reliability question can also work in favor of the foreign steel supplier if disruptions of domestic supply develop. This factor is particularly salient in the reaction of steel consumers when there is a threat of a domestic steelworkers' strike. In this case the riskiness factor reverses; foreign sources of steel supply (along with local non-union sources, if any) become more reliable than domestic sources.

Increased imports may also be the result of local shortages in specific regions not conveniently serviced by domestic firms or of general domestic shortages of specific steel products. Steel, it must be emphasized, is not a homogeneous good. Even in the context of domestic over-capacity, unanticipated surges in demand for specific steel products may leave domestic producers unable to fill orders in a timely manner at any price. The availability of the scarce steel product from foreign sources thereby shifts the competitive advantage abroad by default.

Other non-price characteristics attractive to domestic steel consum-

ers may include quality control and quality standards, delivery time and special customer services. All play a role in the choice of steel suppliers and thereby represent a basis on which domestic producers may lose market share (Comptroller General, 1981, pp. 3.1–3.6.). Thus, both price and non-price factors determine competitiveness; increasing import penetration for any of these reasons may seriously impinge on the domestic steel industry's profitability and challenge it to adjust to the new market conditions.

The economic basis of adjustment
In a setting of unrestricted markets and no government intervention, a trade disturbance will compel the steel industry to adapt to the new circumstances in strict accordance with efficiency criteria. Adjustment thus naturally takes place at the level of the individual steel firm, which combines factors, determines output, and otherwise makes all entre-preneurial decisions based on profit-maximizing motives. In order to retain competitiveness on world markets, a steel firm may, for exam-ple, have to lower its prices, cut costs, retool, adopt new technologies, upgrade quality control, or shift production towards more specialized products. Adjustment may require the steel producer to improve its management practices or to expand or contract its operations. It may require the firm to improve relations with labor unions, or it may require a more aggressive stance by management in wage negotiations. Often, the value of the firm's capital must be written down and in the extreme case, the steel firm may be forced to shut down, either temporarily or permanently, as a result of the trade disturbance. Yet it is crucial in this context to keep in mind that international trade is but one of many sources of economic change that constantly challenge firms to adapt. Countless economic decisions regarding evolving fac-tor, input and product market conditions confront the firm on a daily basis, and require that the appropriate adjustment take place. To single out 'trade' as a special source of disturbance reveals the extent to which the study of economics in general is structured on the existence of political boundaries.

One can, in this regard, view international trade as the great equaliz-er of 'provincial' markets, subjecting all steel firms to the standards of world competition. Even if national markets are not perfectly com-petitive, the introduction of import competition tends to exact from all firms compliance with the rigorous demands of international markets. This insures at once that goods will be produced by the most efficient firms worldwide and that consumers will enjoy the lowest possible prices.

At the same time, the openness of world markets demands a univer-sal, far-sighted scope from firms in their investment and planning

decisions in the medium and long term, when factors are no longer fixed. In a non-interventionist context, risk must therefore be calculated by firms on the basis of present and future world market conditions. It is particularly important that steel firms recognize the nature and implications of changes in world markets in order to adjust in an efficient manner in the long run. Thus the difference between cyclical and structural changes in world demand patterns is crucial to the determination of proper short-run and long-run adjustment, respectively. Even if market developments can in some sense be regarded in retrospect as 'unforeseeable,' a market-driven system demands that the consequences of risk-bearing decisions be internalized as either private gain or private loss by the firm.

Vulnerability to trade disturbances

Despite the dictates of economic efficiency, it is clear that steel firms often seek protection from imports as an alternative to adjustment, and that they often receive it. In this regard, it is useful first of all to identify those factors that make the steel industry particularly *vulnerable* to competitive shocks from abroad. In general, the vulnerability of an industry to trade disturbances depends upon its exposure to competitive pricing and shifting competitive advantage on world markets.

The most direct source of vulnerability lies in the organization of the product market. If the domestic steel industry consists of a monopoly, cartel, or oligopoly, its pricing practices will tend to increase its exposure to import competition. A monopoly, for example, typically equates marginal cost with marginal revenue, such that the resulting price generally lies above the level of marginal cost. Steel export markets, on the other hand, are usually much more competitive, setting export prices much closer to marginal cost. If the domestic market is open to world trade, imports will begin to impinge on the monopolist's market when the marginal cost of foreign exports (including transportation and other delivery costs) falls below the domestic monopoly price. A monopoly or cartel agreement designed to increase industry profits (such as the Eurofer cartel in the EC) must therefore generally depend on import restrictions in order to maintain its desired domestic market power.

This analysis extends to any cartel or oligopolistic arrangement among producers that manages to set prices above marginal cost: every incremental price rise above their more competitive levels on international markets increases the likelihood of market penetration by imports. In national steel industries, where a few large firms tend to dominate the market, traditional pricing practices may leave producers unprepared for marginal cost pricing on world markets. Under

such an oligopolistic market structure, the high fixed costs and fluctuating output:capacity ratios of steel production have often produced a 'cost-plus' pricing mentality, in which firms establish prices as a mark-up over existing average cost. In the absence of vigorous competitive pressures, the oligopoly has less of an incentive to innovate, adopt new technologies and otherwise reduce costs, focusing instead on price discipline as a means of maintaining stability in the domestic market, especially during cyclical downturns in demand. Well-entrenched oligopolistic pricing habits such as these will therefore leave the industry vulnerable to the seemingly severe international pricing practices that emerge from the more competitive export market environment. The oligopolistic United States steel industry, for example, apparently developed a 'cost-plus' pricing tradition and yet remained insulated from import competition for many years by virtue of its superior cost competitiveness. As foreign firms became more competitive in the postwar period, however, United States steel producers became increasingly vulnerable to import competition.

A more general source of vulnerability lies in the economic factors underlying steel production, combined with diverging factor market and other cost conditions among steel-producing countries. Differences in relative factor endowments point to potentially large differences in steel production costs from country to country. New technological developments geared to the more efficient use of a country's abundant factor can create cost advantages there that cause a world-wide shift in steelmaking competitiveness, leading, in turn, to trade disturbances. The cost of constructing new steelmaking capacity may also figure in a country's steelmaking competitiveness once new technologies become available. The combination of new technologies, low labor cost and low steel plant construction cost played a large role in the increasing import market penetration of Japan's steel producers in the 1960s, just as these factors have led to the emergence of some newly industrializing countries such as South Korea as highly competitive producers.

Factor market conditions can, in addition, also combine with product market structure to create cost *disadvantages* for a domestic industry. Under perfectly competitive labor market conditions, for example, one would normally expect wage rates in the steel industry to diverge internationally as the value of marginal product varies. Yet when laborers (or, theoretically, any owners of factors) gain market power and can drive the rate of return to their services above the competitive level, the affected industry will exhibit increased vulnerability to international competition. The existence of sheltered oligopolistic product markets is likely to increase the incidence of monopoly labor practices, since the small number of firms facilitates union organ-

ization and the additional production costs can be readily passed on to the consumer as a cost mark-up. The day of reckoning arrives, as it has in the United States market in particular, when competitively priced imports begin to impinge on the national market, threatening both the oligopolistic pricing structure and the premium wage scale.

Reaction threshold and response

Whatever the vulnerability of the industry, steel firms will not react to a trade disturbance until they foresee import penetration threatening profitability. This reaction threshold is reached when the firm decides that 'something must be done' to counteract the effects of the disturbance. In broad terms, the firm's alternatives are either to adjust according to economic criteria or seek government protection from the disturbance. If government intervention is not available or likely to occur, the firm will have no choice but to adapt to the exigencies of the market, no matter what pain is involved. The very possibility of government-sponsored protection, however, adds a new dimension to the firm's behavior in the face of a competitive shift in world markets.

The reaction threshold for action in response to a competitive shock is thus determined not only by objectively established microeconomic characteristics underlying profit maximization, but also by the perceptions of firms regarding the significance of the disturbance and alternative courses of action. A low threshold level implies a quick reaction by the firm, leading to continuous and smooth adjustment to changing world market conditions. A high threshold level, on the other hand, suggests the probability that adjustment will be delayed or postponed. Central to the determination of the height of the reaction threshold are the concepts of the firm's private adjustment cost and the availability of protection. The higher the firm's adjustment cost and the greater the availability of protection, the higher will be the reaction threshold to a trade disturbance.

The firm's adjustment cost is closely related to the degree of flexibility it enjoys in responding to competitive shifts of any type. High fixed costs resulting from long-term investments and extensive vertical integration, for example, raise the stakes of adjustment. The downward inflexibility of wages creates rigidities in the labor market, with the politically difficult alternative of layoffs. Some firms, especially those in the EC, are in fact subject to government regulation on the extent to which workers can be laid off at all. The specific character of factors of steel production hinders their rapid transfer to alternate uses. Those elements of the production process, in other words, that tend to increase the private cost of steel firms to adapt to changes in markets contribute to the firm's reluctance to adjust.

The scope of operations of the firm and the international structure of

production represent other possible indicators of adjustment flexibility. A highly disaggregated, mobile production process would point to the option of transferring certain manufacturing stages abroad, especially if factor cost advantages are involved, thus opening a new path of adjustment. Many consumer electronics industries have adjusted to competitive shocks in this manner (Walter and Jones, 1981). In contrast, a rigidly structured production process such as steelmaking, which relies on the close physical proximity of the various stages, would suggest severe limitations on the firm in dealing with shifts in cost advantages. In general, one would expect transnationally organized production processes to afford a greater degree of adjustment flexibility than those tied to individual countries or regions.

In the context of these constraints and firm characteristics, the decision taken by steel firms faced with a trade disturbance will be the result of a consideration of the costs and benefits of adjustment. The financial cost of adjustment is compared with the cost of *non-adjustment*; presumably, firms are rational economic actors and will choose the path of lowest cost. Yet, as in all business decision-making, there is an element of uncertainty associated with the consequences of a trade disturbance. The cost of adjustment can be calculated by the firm using the requisite market information to determine expected price trends and the optimal manner in which to physically alter the plant and equipment, adjust the factors of production, etc., in order to remain competitive. If outcomes are purely market-determined, the expected cost of non-adjustment will be calculated as the stream of expected revenue losses over time attributable to the lack of adjustment. The exact present value of these losses will depend upon the course the market takes, and the firm's management must therefore identify the nature and duration of the disturbance in order to properly estimate the stakes of adjustment. If a long-term trend in demand or supply is mistaken for a short-term trend, for example, the stakes of adjustment will be underestimated.

Therein lies the crucial determinant of the height of the reaction threshold. If all firms were blessed with perfect market foresight and acted upon it, there would be no need to distinguish between those subject to high, as opposed to low, adjustment costs, since the reaction threshold would be determined solely by the point where the present value cost of non-adjustment exceeds that of adjustment. However, it is conceivable, and even probable, that the existence of high adjustment costs will induce an over-optimistic attitude towards the trade disturbance, a feeling that 'the trouble will pass,' leading to an underestimation of the cost of non-adjustment. The determining factor in this regard is the firm's ability to recognize and evaluate the implications of trade-related disturbances, which is a function of the past

experience of the firm and the perceptions, not to mention the pers-
picacity, of its managers. A steel industry untouched by import
penetration for many years (the US industry until 1959, for example),
may be particularly vulnerable to trade disturbances, which demand
from entrepreneurs the ability to look beyond national markets and
judge trends in competition worldwide. The increasing involvement of
government in the steel industries of Europe in the 1960s and 1970s,
whose goals were tied largely to social and regional policies, tended to
cloud the entrepreneurial vision needed to remain competitive in a
rapidly changing world steel market. Finally, the high cost of adjustment may present management with the prospect of some unpleasant
decisions, such as shutting down an obsolete plant, laying off workers,
or writing down the value of the firm's capital. These factors all
contribute to the under-estimation of the cost of non-adjustment,
which creates a high reaction threshold to international competitive
shifts.

The high threshold barrier has, moreover, a built-in dynamic resistance to adjustment, since an initial reluctance to adapt to a trade
disturbance tends to create a vicious circle of competitive decline.
When import penetration begins to impinge upon a domestic firm's
revenues, for example, the firm's refusal to adjust exacerbates its
competitive position and impairs its ability to provide resources for
adjustment in the next period. As import penetration continues and
the need to adjust becomes more urgent, the requisite adjustment
measures become more painful, leading, by the analysis above, to an
even higher reaction threshold.

It is unlikely that a firm would allow its position to deteriorate so far
without the existence of an alternative to unrelenting market forces as
the means of retaining its viability, an instance, in other words, of last
resort in dealing with the trade disturbance. The firm's expectations
regarding government trade protection (or other intervention to aid
the industry) therefore tend to raise the level of the reaction threshold,
since the expected cost of non-adjustment is reduced.

The anticipation of non-market forces introduces a new dimension
into the analysis of adjustment. The discussion so far has assumed that
the risk associated with entrepreneurial decisions is internalized as
private gain or loss. It is clear, however, that the possible socialization
of risk by state intervention will change the firm's otherwise market-
driven behavior regarding adjustment. One would expect that firms
faced with higher adjustment costs are more likely to seek protection
than those with low costs. The denial of the market forces of adjustment is caused by the fact that firms can also influence and anticipate
policy decisions that affect them and thereby supplant the normal
microeconomic criteria for profit-maximization. An initial failure to

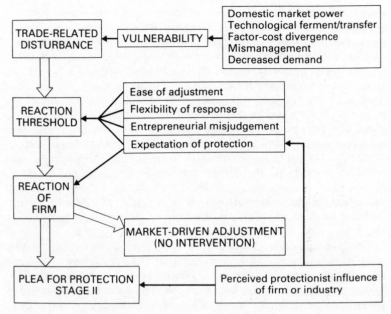

Figure 3.1 *The creation of protectionist pressure.*

adjust may in fact drive the firm to seek influence within the government to secure a pledge of 'insurance' from import disruption for possible use in the future. The firm's optimism regarding future government action will in the meantime feed back to influence the reaction threshold level. The reaction threshold thus rises, the more firms perceive their own political influence to be increasing or the more they perceive government institutions and policymakers to be prone to grant them protection.

Summary of stage I

Stage I of the course of generating steel protectionism can in general be described as an economic process that follows the impact of a trade disturbance. Figure 3.1 illustrates this process schematically. A set of production and market characteristics contribute to the vulnerability of the industry to trade disturbances. When the disturbance actually begins to reduce the firm's profitability, firms seek channels of adjustment to adapt to the situation and return to previous levels of profitability. The availability of low-cost adjustment channels, found in disaggregated aspects of the production process and firm organization, points to a high degree of flexibility of response and low adjustment cost, thereby creating the low reaction threshold characteristic of

market driven adjustment. In contrast, a low flexibility of response, linked with high fixed costs, points to high private adjustment costs and a reluctance to use the available economic adjustment channels. The resulting high threshold barrier delays adjustment and causes the firm to look more towards government intervention as an alternative. Meanwhile, expectations regarding protection affect the reaction threshold. The ultimate lowering of the reaction threshold induces adjustment to the disturbance and gives the story an economic ending, with no government intervention. A chronically high reaction threshold, on the other hand, moves the story to the political sphere, where a debate over the granting of protection takes place.

The generation of steel protectionism

The economics of adjustment and non-adjustment in the steel industry is linked to commercial policy by a system of competing influences acting within a framework of government institutional structures. In stage I, where the firm initially responds to a trade disturbance, economic forces dominate, but in stage II, where the steel industry attempts to secure protection, political forces dominate. The pivotal element in the abandonment of market-driven behavior in this context is the tendency for political expectations to disrupt economic processes. When various economic factors affected by the trade disturbance, such as the steel industry executives, stockholders and labor unions, see greater expected benefits in seeking protection than in economic adjustment, consistent rational economic behavior will cause them to choose the former. Their efforts are thus rechanneled towards protectionist advocacy in the political debate that follows. The outcome of the debate, as will be shown in the following discussion, depends on the relative strengths of interacting pro- and anti-protectionist forces.

The advocates of protection

It was established in the previous section that the private cost of adjusting to a trade disturbance can be very high, and that the higher it is, the more reluctant the firm will be to adjust, and the more likely it will be for the firm to seek protection. The cost of adjustment does in fact define the motivation and identity of proponents of protection in the private sector of the economy. For example, a long-term decline in the price of steel may force the firm to write down the value of its capital in order to remain competitive. Similarly, the firm may find that expensive investment in new technologies is required in order to retain viability in the marketplace. The cost of these decisions must be borne ultimately by the firm's stockholders. Thus, the owners of capital bear

the brunt of the private cost of adjustment to a competitive shock when the only available economic channels of adjustment involve a large decrease in the book value of the firm's capital stock. Under these conditions one would also expect a decrease in the traded price of the steel firm's stock. Stockholders may relinquish their holdings of the stock (probably at a loss); otherwise their interests lie in protectionist policies in so far as these are capable of protecting the book value of the firm's capital.

Similarly, steel executives can also develop a strong reluctance to adjust and will often figure prominently in a protectionist campaign. As those in charge of the operations of the firm, their decision-making ability will play a large role in determining its competitive position on the world market. To the extent that steel executives can be held reponsible for riskbearing decisions, they will be exposed to blame for the failure of the firm to adjust. Specifically, the owners of the firm generally expect the firm's management to insure that an adjustment path of lowest cost is followed; managerial decisions that lead to costly delays in adjustment, or that otherwise raise the private cost of adjustment, place the careers of those responsible in jeopardy. It is therefore in management's interest to see the blame for their fateful decisions (or indecision) shifted elsewhere. Government intervention in the form of trade restrictions serves this interest; it absorbs the consequences of mismanagement by shifting the cost to consumers and tends to direct the blame for the industry's situation towards 'disruptive' foreign suppliers.

The position of labor in the steel industry may also be to resist adjustment, although its interests are not so clear-cut as those of stockholders and management. If comparative advantage in the importing country is shifting in such a way that laid-off steelworkers can be absorbed by industries offering a higher wage, then the workers themselves will tend to endorse – indeed, embrace – adjustment. Theoretically, adjustment in accordance with comparative advantage will in any case cause total national welfare to increase, thus allowing the use of the compensation principle to make workers better off afterwards (Stolper and Samuelson, 1941).

However, there are likely to be other complicating factors at work that mitigate against this sanguine prospect, especially when workers are members of powerful unions in industries located in areas lacking alternative employment opportunities. The possibility was mentioned earlier that monopoly labor practices could actually help to bring on the trade disturbance. If in fact the workers' wage is that much above the value of marginal product, adjustment to the disturbance would probably lead to the loss of the union's market power. A strong incentive is thereby created for them to resist adjustment. Even when

the wage they receive does not exceed the free market level, union workers may value the additional features of unionization (collective bargaining, right to strike, etc.) more than the wage differential without union benefits they might receive at a new job. One must add that labor union leaders are in any case likely to resist adjustment when it diminishes the rank-and-file, since the interests of union leaders lie in maintaining an organization as large and powerful as possible.

Resistance to adjustment by workers can also take the form of regional attachments or a resistance to change in general. If other employment opportunities in the region are scarce, laid-off workers often face the need to move their residence in order to find new employment and maintain or improve their families' economic welfare. The ultimate reason for their reluctance may lie in a certain fear of the unknown: a new job or trade, new employer, new environment, etc. To be sure, such fears cannot be taken for granted. Economic incentives are generally a powerful force in fostering adaptive responses to change. Economic expansion in the United States, for instance, has historically gone hand-in-hand with shifts in population centers and sectoral employment patterns. One must instead view resistance to adjustment in terms of the workers' perceptions of both the cost of adjustment (financial and psychological) and the possible alternatives. The very possibility of government intervention tends to raise barriers to change, since a viable alternative to change is put into view. In short, workers will tend to resist adjustment when the expected net benefits of government intervention and non-adjustment exceed the expected net benefits resulting from adjustment.

The political context
The receptiveness of government officials to these groups' plea for protection depends on an often subtle formation of public and official attitudes based on the economic, political and institutional environment in which policymaking takes place. These are the elements that magnify or diminish the political power of both protectionist sentiment and efforts to maintain open trade policies. The outcome of the debate between the forces for and against trade restrictions will be the result of the relative strengths of each side. One can characterize the debate in structural terms by picturing a barrier of governmental reluctance to implement trade restrictions that protectionist influence must surmount in order to move the government to action. The strength of the forces aimed at the barrier, as well as the height of the barrier itself, is strongly affected by the policy environment, as mentioned above. The various interacting components of this system of policy formation are the subjects of the following discussion.

The advocates of protection, in their efforts to effect the imple-

mentation of trade-restricting policies, attempt to transform their reluctance to adjust into positively directed political influence on government authorities. In a representative democracy, the most sensitive 'pressure points' are found among elected officials, where constituency support is most often crucial to political fortunes. The most direct means of realizing political influence are lobbying groups charged with effecting passage or blockage of specific legislation, but broader-based attempts at achieving 'grass roots' support through propaganda and demonstrations work to the same end. The amount of resources used in a protectionist campaign depends in general on the stakes involved, which in economic terms is the expected transfer to the domestic industry through the price and output increases caused by the trade restriction. At the same time, the available resources and degree of organization of the protection-seeking groups impose constraints on their influence. It is therefore not surprising that the most politically influential lobbies in western industrial democracies include steel corporations, with geographically concentrated production and large, nationally organized labor unions.

The sectors of the economy that would benefit from open trade in steel are, in contrast, much more dispersed and therefore wield much less political influence. The fact that the benefits from free trade are thinly spread while the losses from it are highly concentrated thus profoundly influences the political economy of trade restrictions in representative democracies. To be sure, some groups have a substantial interest in open trade policies, especially steel importers and steel-using industries that compete on world markets, as well as exporters in the steel-importing country who would suffer from foreign retaliatory protectionism or other ill effects of restricted trade (overvaluation of the exchange rate, decreased foreign income that reduces market opportunities, etc.). The political influence of steel importers and steel-using industries pales, however, in comparison with the heavily concentrated and well-organized domestic steel lobby. For exporting industries, the losses due to steel protectionism may be indirect or difficult to identify, diminishing the emotional appeal and conviction with which they may oppose the trade restriction. Moreover, those that gain the most in aggregate terms from open trade in steel, domestic consumers of products made of steel, are difficult to organize politically because of their geographic dispersion and the fact that the benefits are small per consumer.

It is therefore not the structure of political influence, but rather a general commitment to free trade principles by policymakers that represents the interests of consumers. In so far as policymakers give priority to aggregate national economic welfare over the sectoral concerns of the domestic steel industry (including its formidable trap-

pings of national prestige and well-being), the pursuit of open steel trade becomes a rational policy objective. Freer trade increases aggregate welfare; protectionist policies not only diminish it in a static sense, but are also likely to lead to further losses in welfare through retaliation abroad and a further deterioration in steelmaking and general economic competitiveness on world markets.

If protectionist policies are not already in force, the barrier of government reluctance to impose trade restrictions will generally be stronger due to the typical bureaucratic resistance to change. Protectionist lobbies, like any other interest group, set out to force government officials to implement a new policy that will benefit them. These efforts automatically challenge the *status quo*, exposing the targeted officials to countervailing pressures, either from domestic opponents to protection or from the foreign governments that will suffer from it. Protectionist policies force governments into positions of compromise, and the political stakes of protection must be large enough to justify the efforts needed to appease its opponents, especially foreign governments. In addition, domestic government officials know that by granting protection once, they may be faced with even more requests for protection. For these reasons it is important for the protectionist lobby to cause its interests to be internalized by those government officials responsible for the implementation of trade restrictions.

The government's commitment to the principles of open trade policies, plus the inherent bureaucratic resistance to change, establishes the barrier to the realization of trade restrictions that protectionist forces attempt to surmount. Interest groups thus try to penetrate the policymakers' decision-making criteria in order to gain sympathy for their cause. Conditions in the political and economic environment, as well as the structure of policymaking institutions, will determine the degree to which policymakers are open to protectionist influence. In structural terms the creation of *resonance* in the policymaking structure with protectionist interests leads to a strengthening of their forces to a point closer to the requisite threshold level of influence needed to receive protection.

The formation of public and official attitudes towards protection are shaped in part by the condition of the economy as a whole. High unemployment and slow growth tend to create a bias against imports, since government officials may perceive adjustment-induced labor displacement and reallocation of resources involving plant closings as a signal of economic weakness requiring intervention. The traditional association of steel with economic well-being amplifies these concerns. A high rate of inflation, on the other hand, tends to dampen the bias against imports, since they have a moderating influence on the general price level. Yet serious levels of macroeconomic distress in general

may also foster xenophobia, isolationism and a general state of discontent in society, all of which could contribute to protectionist sentiment. The overall social mood of the country provides a channel through which protectionist sentiment can achieve resonance among policymakers.

This channel of influence becomes more direct as policymakers perceive the fate of the steel industry as being crucial to the achievement of certain national goals of high priority, such as employment, national defense, the balance of payments, or independence from foreign suppliers. The fulfilment of these goals involves a cost to society, and policymakers will ideally equate their marginal social value to their marginal social cost and consider alternative instruments in pursuing them (see Corden, 1974, Chapters 2 and 3). Yet some elected or appointed officials may in addition associate the fate of the steel industry with more narrowly defined goals, such as regional development, which may have an impact on their political careers. Others may harbor a genuinely interventionist ideology, inherently distrusting the international market or fearing the disruptive effects it might produce.

The inner dynamics of legislative and administrative institutions can also contribute to the responsiveness of government to calls for protection. In the legislature, the lack of party discipline, party leadership, or a well articulated party platform regarding trade restrictions promotes logrolling and an *ad hoc* accommodation of lobby interests that favor protectionism. In administrative bodies charged with enforcing trade restrictions or regulating steel industry activities, protectionism may have automatic resonance with policymakers. In fact, such bureaucratic bodies may have a vested interest in protection if it appears necessary to prevent the failure of previous policies carried out under their direction. In the United States and the EC, the steel industry's chronic, protracted distress thus creates a sort of interventionist dynamic: the failure of one policy to aid the industry tends to produce more elaborate policies, which inevitably include increasingly restrictive trade policies.

Summary of stage II

Figure 3.2 illustrates the process of generating steel protectionism. The industry's protectionist campaign attempts to wield political power to surmount the barrier to the implementation of trade restrictions erected by open trade influences in the government's policymaking structure. The government's reluctance to protect is based variously on the expected negative welfare effects of protection and on the additional economic and diplomatic effects of foreign retaliation. The political influence of the protection-seeking steel industry is based on

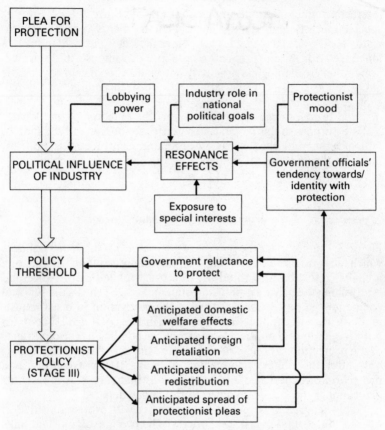

Figure 3.2   *The process of generating protectionist policy.*

both direct and indirect means. The direct means of industry influence lie in its lobbying power and its ability to directly pressure legislators. Indirect means include influence gained through the protectionist mood of the country, the perceived importance of steel to specific political goals of the government, and the exposure of policymaking organs to special interests, all of which allow protectionist sentiment to resonate within the government and thereby penetrate the policymaking process. As a result, the government's tilt towards protectionism, which was originally limited by the interventionist mentality of government officials themselves, increases, strengthening protectionist influence. When the degree of influence surpasses the requisite threshold level, a protectionist policy will be adopted.

The generation of protectionism thus involves competition between

interacting influences and anticipated effects. Since the debate pits special interests against more general commitments and attitudes towards open trade, it is the degree of interaction of the steel industry's influence with the government's propensity to impose trade protection, either by ideological proclivity or organizational capacity, that is likely to be decisive. This structural approach cannot, of course, capture all aspects of the complex nature of protectionist policymaking, and therefore cannot always predict the outcome of the debate. Yet the purpose of this model is to identify the sources and channels of influence that underlie the process of generating protectionist policies. Shifts in the patterns of influence and in the access to channels of resonance and response point to shifts in policy trends.

A perspective of the protectionist policymaking process

Figure 3.3 summarizes the process of generating protectionist policies, which falls into three stages. Stage I is essentially economic, presenting a challenge to the firm to adjust to a trade-related disturbance and establishing the private cost of adjustment. Stage II shows the political component of protectionism, tracing the motivation to resist adjustment through the application of political influence. Stage III contains both economic and political elements corresponding to various effects of the protectionist policy. In so far as the impact of protection occurs only after the policy is implemented, stage III exists purely as a projection of experience and expectations that serves as a basis for determining the stakes involved in the policy debate.

Yet, in general, the expectations associated with protectionism form the most important part of the process. The two most important links in the concatenation of events that lead to protection occur between stages I and II, where market-driven adjustment is abandoned in favor of a plea for protection, and between stages II and III, where the protectionist policy threshold is surpassed by protectionist influence. The direction of the entire process hinges on the expectations formed at these two points regarding the magnitude of the benefits that the steel industry will receive from trade protection and the probability that the government will actually grant it. The benefits or stakes of protection are estimated by a prior determination of the effects of income redistribution that will benefit the industry at stage III. These effects form the basis for the political incentive structure that causes the government to consider granting protection; they thus feed back to stage II, where the government's propensity to protect is determined. At stage II, factors within the political environment can enhance the likelihood of government intervention. Information regarding the net

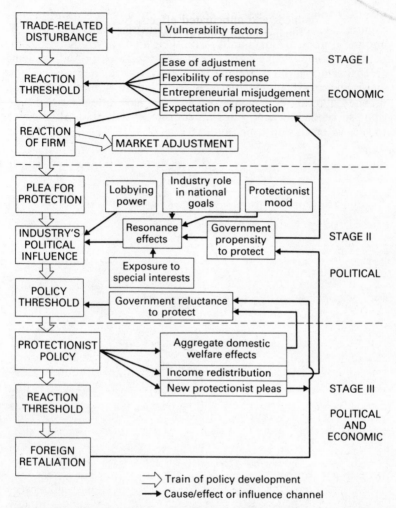

Figure 3.3   *The three stages of protectionist policymaking.*

expected benefits then feeds back to stage I, where the anticipation of protection determines the path taken by firms faced with the need to adjust to the disturbance. A heightened expectation of government protection is thereby readily transmitted to the stage of the steel firm's adjustment decisions and allows the generation of trade restrictions to begin. Once the industry estimates the stakes of protection, it can then proceed to devote a commensurate amount of its resources to the political effort to achieve protection. Political expectations that sup-

plant economic processes thus constitute the driving force of protectionism in general. The integrated steel industries of the United States and the European Community, with their high vulnerability to competitive change, strong resistance to the private cost of adjustment, and expectations of protection from imports (based on experience and increasing government involvement in the industry), have persistently rejected market adjustment. Instead, they have often pursued, and often succeeded, in achieving the requisite level of political influence to receive protection from imports.

4

The World Steel Market in Transition

The analysis of the previous chapter showed that protectionism results from the rejection of market-driven adjustment to increased import competition or the threat of it. In the steel industries of advanced industrialized areas, especially the United States and the EC, increased steel imports since the late 1960s have led to a continual stream of protectionist activity. This chapter sets out to identify the major causes of the changing competitive structure of world steel trade that led to the adjustment crisis of the postwar period. The analysis will focus on the changing economic factors that determine the worldwide pattern of steelmaking competitiveness, such as trends in steel supply and demand, production technology, the cost of inputs and managerial performance. The resulting picture provides some insight into the current pattern of comparative advantage and the trade patterns that would occur under an open trading system. Notwithstanding the increasing involvement and intervention of governments in steel markets, an orderly and lasting resolution of steel trade disputes will depend on a recognition of the economic forces at work in the world steel market.

The chapter begins with a discussion of trends in import penetration in the postwar period and their role in creating protectionist policies. There follows an examination of the sources of structural change in the world steel market and a review of the evidence on the impact of government involvement in the steel industry on steel trade. The chapter concludes with a summary of the role of world steel trade in the adjustment crisis in the established steel industries of the United States and the EC.

Import penetration and protectionist response

Steel trade in the postwar period became a political issue in the United States and the EC when imports began to impinge on domestic mar-

kets and industry profitability was in decline. Although an examination of the trends in steel technology, demand and cost factors casts doubt upon the alleged culpability of trade, *per se*, in the plight of declining steel firms, a review of trade activity reveals the political sensitivity of steel imports. Figure 4.1 shows import penetration levels, steel employment levels, steel price trends and trends in US steel industry profits as a percentage of shareholders' equity compared with the manufacturing average in the United Sates from 1959 to 1984. Sharply increased import penetration during the 1960s, combined with sub-par profit performance of the industry throughout the period, led to the VER agreements with Japan and the EC in 1968. Protectionist sentiment declined from 1972 to 1975 as the industry enjoyed a steel boom, with profitability approaching or exceeding the manufacturing average. A drop in profitability and an increase in imports led to the antidumping suits of 1977 and the beginning of the trigger price mechanism in 1978. The increased import penetration over the period 1980–4, accompanied by a severe deterioration in the industry's financial performance, led to the US–EC steel pact of 1982 and the VER agreements of 1984. One must add that the profitability statistics actually underestimate the losses of the large integrated steel firms, especially in the latter portion of this period, since they incorporate the profits of mini-mill producers, an increasingly important subsector of the US industry (to be discussed below) that has consistently outperformed the integrated firms.

Figure 4.2 shows import penetration levels, steel employment levels and pre-tax profits per ton shipped in the EC from 1967 to 1984. Although import penetration from third countries has not been as high in the EC as in the United States, small increases have been enough to trigger protectionist responses. Increasing imports in the late 1960s and declining profits led to the first VER agreement with Japan in 1970; a similar situation with a steep drop in profits resulted in a more comprehensive VER with Japan in 1975. The nearly continual red ink since that time has, since 1977, led to a strict system of nearly global VER coverage as part of the Community's steel policies.

The sources of competitive decline

Protectionist trade policies in the United States and the EC have often assumed a cause-and-effect relationship between steel imports and a deterioration in the industry's performance, implying that trade restrictions will contribute to the recovery of the industry. However, a broader examination of the industry since the Second World War reveals a worldwide market that has changed radically over a period of

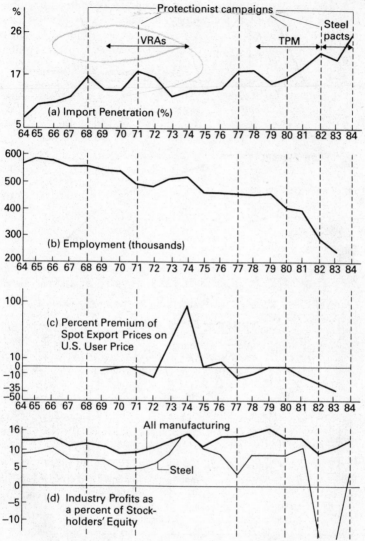

Figure 4.1   *Protectionist indicators in the US steel industry, 1964–84.*

Sources: A, B. AISI, *Annual Statistical Report*, various issues.
C. *World Steel Dynamics*, 'Steel Production and Trade Outlook',
30 November, 1984 (New York: Paine Webber).
D. *Quarterly Financial Report*, various issues (Washington, D. C.: Federal
Trade Commission and Department of Commerce).

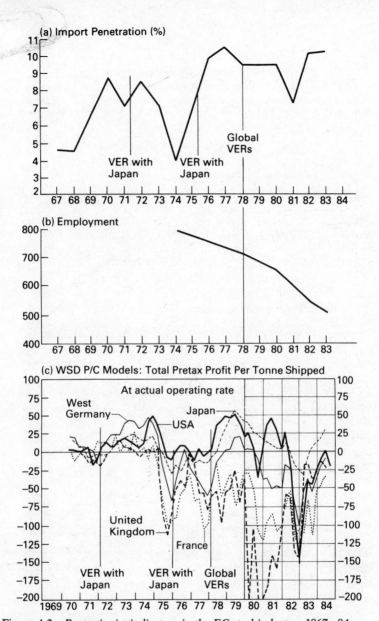

Figure 4.2 *Protectionist indicators in the EC steel industry, 1967–84*.

Sources: A. *Iron and Steel Yearbook*, various issues (Brussels: Statistical Office of the European Communities).

B. *The Steel Market in 1983 and the Outlook for 1984* (Paris: Organization for Economic Cooperation and Development, 1984).

C. *World Steel Dynamics*, 'Steel Production and Trade Outlook', 30 November, 1984 (New York: Paine Webber).

30 years, leaving many of the established integrated steel firms in these countries in competitive decline. In this context imports have most often represented a symptom rather than a cause of decline in the industry. In focusing policy measures on import levels rather than the root causes of the adjustment problem, however, protectionism has only tended to postpone a solution and perpetuate the crisis.

The changing volume and sources of steel trade
An overview of world steel production and trade in the postwar period reveals a dramatic increase in the volume of steel production and exports and a dramatic shift in the sources of exports, factors that have contributed heavily to protectionist policymaking. In the early postwar period, the United States, whose productive capacity was untouched by the war, stood as the focus of steel production and trade. Of the 136 million tons of worldwide steel output in 1947, the United States produced 77 million and exported 4.2 million, making the American steel industry the leading world exporter as well as producer. Among the few other countries producing steel at all that year, only Great Britain, the Soviet Union, France, Belgium and West Germany had steel output of more than 1 million tons. For countries recovering from the war, the bulk of steel output was consumed domestically, leaving little for export (Hogan, 1983, p. 192).

As the postwar reconstruction of industrial capacity continued, the volume of world production and trade also grew (see Figure 4.3). Raw steel production rose to 192 million tons in 1950, while exports rose to 20.5 million tons. By 1960, production and trade figures had risen precipitously to 374 million and 57.2 million tons, respectively. Despite the increased volume of steel trade during the intervening decade, however, steel imports had still not become a major political issue, thanks largely to the continued expansion of the world economy in general and of steel demand in particular.

As the growth of world steel output and trade accelerated from 1960 to 1979, however, imports of steel began to impinge on domestic markets of major producers, as shown in the previous section. The roots of this phenomenon lay in part in the continued expansion of steel production in certain countries beyond the level of domestic consumption. For countries that enjoyed a competitive advantage in various steel products on world markets, this was a natural phenomenon, since export market sales allowed producers to achieve economies of scale in large integrated plants. In addition, a process of specialization within the world steel industry was emerging, whereby large-scale production of certain types of steel in a country or region led to increased exports of those steel products and imports of others, creating a pattern of intra-industry trade.

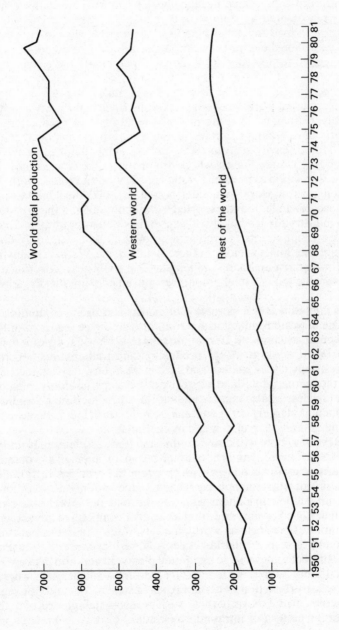

Figure 4.3 *World crude steel production, 1950–81 (million metric tonnes)*.

Table 4.1   *Country shares in the world export market of semi-finished and finished steel, selected years 1965–82*

Exporting countries	1965	1970	1974	1975	1976	1977	1978	1979	1980	1981	1982
Japan	16.1	19.9	25.6	26.3	29.4	29.0	24.9	24.3	24.2	22.8	25.7
European communtity[a]	49.5	41.8	46.0	42.6	37.3	41.5	44.3	44.4	44.6	44.8	41.5
United Kingdom	6.5	4.7	2.7	2.9	3.0	3.8	3.5	3.6	2.3	3.2	3.2
United States	3.8	7.3	4.2	2.5	2.0	1.6	1.9	2.1	3.1	2.2	1.6
Soviet Bloc	17.1	16.9	12.9	15.8	15.0	9.6	9.3	9.1	9.7	11.0	10.2
Other	7.0	9.4	8.6	9.9	13.1	14.5	16.1	16.5	16.1	16.0	17.8

Source: *Statistics of World Trade in Steel* (Geneva: United Nations Economic Commission for Europe, 1978)

[a] The European Community here refers to the original six member countries – France, the Federal Republic of Germany, Italy, Belgium, the Netherlands and Luxembourg.

The United States had progressively decreased its participation in export markets throughout the 1950s, an early indication of competitive decline, and became a net importer of steel in 1959 for the first time since 1897. The countries of the EC, another traditional steel exporting area before the Second World War, re-emerged as major steel exporters in the 1950s, although their share of the world steel export market declined after 1960 (see Table 4.1). The most significant development in steel export markets in the postwar period, however, was the emergence of Japan as the leading single exporter of steel and of several newly industrializing countries (NICs) as a major factor in world steel trade. Japan, as the following analysis will show, became the most cost-efficient large-scale producer and therefore a major target of protectionist campaigns and policies. The NICs, while they have not supplanted the major steel exporting countries, have nonetheless become competitive in several markets, increasing exports rapidly since the mid-1970s (see Table 4.2). They have therefore also found themselves the target of trade restrictions.

The dramatic increase in world steel production, which grew by nearly sixfold from 1947 to 1979, combined with the increasingly dispersed structure of the export market, gave rise to a severe adjustment crisis in the American and EC steel industries. Part of the problem lay in what was described as 'excess capacity.' As more and

Table 4.2 *Steel exports of newly industrialized countries, 1973–9 (millions of metric tonnes)*

Country	1973	1974	1975	1976	1977	1978	1979
Korea	962	1,329	1,002	1,399	1,389	1,822	3,000
Taiwan	249	194	238	280	312	892	1,038
India	150	186	348	1,522	1,312	500	200
Australia	1,396	1,238	1,727	2,224	2,482	2,573	2,000
Canada	1,273	1,469	1,051	1,518	1,717	2,738	2,827
Mexico	141	122	76	154	275	350	400
Brazil	431	236	149	262	364	936	1,385
Argentina	625	380	54	356	280	743	450
Spain	1,712	800	1,580	2,443	2,678	4,117	4,100
South Africa	632	612	324	1,085	2,096	2,215	2,600
Total	7,571	6,574	6,549	11,243	12,095	16,886	18,000

Source: UN, study by Japanese steel export manufacturing committee; see Hogan, 1983, p. 86

more countries expanded their steelmaking capacity beyond domestic demand, the number of available export markets began to shrink. While the resulting problem of persistent gluts on world export markets at first appeared to be cyclical in nature, it became clear by the late 1970s that it was essentially structural. The solution to the particular problem clearly lay in a reduction in world steel capacity, but here is where the political problems began. Where should the cutbacks take place? Market-driven adjustment would focus the reductions on ageing capacity in the United States and the EC, but the associated political pain and social disruption have played a large role in the resistance to the adjustment process in these countries. Trade restrictions have emerged as the principal means of delaying the retirement of uncompetitive steelmaking capacity.

The secular decline in steel demand
While the problem of excess capacity has thus far been described in terms of the large increases in steelmaking capacity and supply, it is equally a problem of declining demand. One of the major problems facing any steel producer is that of aligning productive capacity to expected levels of demand over the business cycle. As was noted in Chapter 1, the demand for steel fluctuates radically between boom periods and slumps. Yet in planning for an optimal (that is, cost-effective) level of steelmaking capacity over the business cycle and beyond, steel firms must anticipate not only current trends in their competitors' production and in industry demand, but also long-term

trends in the demand for their products. A particularly serious problem occurs when a long-term, secular decline in the demand for steel is mistaken for a short-term fluctuation or vice versa. The decade of the 1970s posed such a problem to steelmakers worldwide, who had difficulty in separating short-term from underlying long-term trends in the steel market. The boom period of 1973—4 was one of widespread optimism in the industry and generated sanguine forecasts of expanding steel markets for the foreseeable future. The collapse of steel demand in the mid-1970s, on the other hand, was widely regarded as a temporary phenomenon and did not prevent steel capacity from increasing apace, especially in Europe.

These short-term phenomena masked, however, an overall, long-term decline in steel demand in the industrialized countries. A number of interrelated factors were converging at this point to amplify the decline. First of all, overall steel *intensity* among the major industrialized steel producers was in decline. Steel intensity refers to the ratio of apparent steel consumption to GNP. As Figure 4.4 shows, steel intensity appears to rise as GNP per capita rises at the lower end of the scale, but then begins to decline past a certain level of GNP per capita (about $2,000 per capita in 1963 dollars). The steel intensity curve suggests that one aspect of comparative advantage in international steel trade, domestic demand patterns, is shifting away from advanced industrialized countries in favor of newly industrialized countries. The economic reasoning behind this idea is based on an apparent migration of steel consumption patterns during the course of economic development.

This relationship implies that, as industrial economies mature, their consumption of steel tends to decline, suggesting a sort of 'stages of growth' model of steel usage. At very low levels of per capita income, before industrialization has begun, an economy typically consumes a correspondingly small amount of steel. Most less-developed countries fall into this category. As an economy begins to industrialize, steel consumption tends to rise more quickly than per capita income as economic activity focuses on building infrastructure (railroads, bridges, communications facilities, etc.). European countries during the postwar recovery (until the 1960s) and, currently, the newly industrializing countries, such as Brazil and South Korea, would be included in this group. As industrialization proceeds, a third stage is reached, in which the growth rate of steel consumption matches the growth rate of the economy. The United States exhibited this relationship from the mid-1950s to the mid-1960s. Finally, as the economy matures further, economic activity shifts progressively away from steel-using and other manufacturing industries and towards service-related industries, leading to a declining steel intensity curve. Such post industrial economies

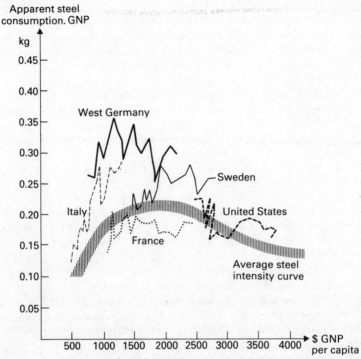

Figure 4.4   *Steel consumption intensity, industrialized countries: yearly fluc-tuations in the apparent steel consumption to GNP ratio, 1957–74 (GNP at constant 1963 dollars).*

now include the United States, most of western Europe and Japan.

Technological and manufacturing design factors have amplified the overall decline in steel comsumption in the advanced industrialized countries. The development of stronger, lighter specialty steels and the increased efficiency of steel usage in manufacturing have meant that the same (or greater) economic utility comes from lower tonnages of steel output. This trend is in part a result of competitive pressures to improve the energy-efficiency of goods such as automobiles by reducing their weight. In addition, these countries are shifting their industrial output more and more towards goods that benefit from advances in specialty steel, such as high-technology equipment, machinery and aircraft, which implies that their competitive advantage is shifting away from large-volume carbon steel production and towards lower-volume specialty steel production.

Fundamental shifts in the competitiveness of steel-using industries

have also had repercussions on the domestic demand for steel. Traditionally heavy consumers of steel in the industrialized countries, such as the automobile and shipbuilding industries, have suffered from competitive decline on international markets. When their output falls, the derived demand for domestic steel tends to fall as well. The indirect importation of steel in the form of automobiles and other goods with a large steel content, in particular, has eroded the domestic steel markets of the United States and several European countries (Keeling, 1982, pp. 10–18).

Other technological advances have created substitutes for steel. For example, plastics and aluminum have partially replaced steel in automobile production, plastic and concrete tubes have replaced steel tubes in some cases, and steel-reinforced concrete has replaced structural steelwork in construction. At the same time, other developments have worked in the opposite direction. Steel has replaced glass in much of the beverage container industry, while it has displaced timber, brick and concrete in construction (Keeling, 1982, pp. 15–17). Overall, however, studies indicate that technological factors have been reducing steel intensity at a rate of 0.5 per cent per year (Keeling, 1976, p. 17).

In general, the traditional steelmaking countries of the world thus face gradually shrinking markets over the long term. While economic booms may temporarily increase steel consumption above the trend line, the overall intensity of steel demand in the United States, Europe and even Japan has fallen steadily over the past decade and is likely to continue in this direction.

Technological competitiveness

As steel demand has declined in most advanced industrialized countries, technological adaptation has become a crucial factor in maintaining competitiveness. Although a comprehensive review of technological developments goes beyond the scope of this study (see OTA, 1980; Barnett and Schorsch, 1983), two major factors in steelmaking competitiveness will be examined here: furnace technology and continuous casting. Differential rates of adaptation in these areas have played a large role in changing the pattern of competitive advantage on world markets.

There are essentially two types of efficient steel 'melt' technologies available to steelmakers today, the basic oxygen furnace (BOF) and the electric arc furnace. The BOF has replaced the open hearth furnace as the principle method of large-scale steel production. Developed in Switzerland and Austria, BOF technology was first put into operation by the Austrian firm VOEST in 1952 and is widely regarded as the most important innovation in steelmaking in modern times

Table 4.3 *Adoption of efficient steelmaking technologies, various countries (percentage share and millions of net tons)*

	USA		Japan		EEC(9)	
	%	tons	%	tons	%	tons
BOF						
1960	3.4	3.3	11.9	2.9	1.6	1.8
1965	17.4	22.9	55.0	24.9	19.4	24.3
1970	48.1	63.3	79.1	81.2	42.9	65.1
1975	61.6	71.8	82.5	92.9	63.3	87.2
1981	60.6	73.2	75.2	84.1	75.1	103.6
BOF plus electric furnace						
1960	11.8	11.7	32.0	7.1	11.5	12.4
1965	27.9	36.7	75.3	34.1	31.5	39.5
1970	63.5	83.5	95.9	98.4	57.7	87.6
1975	81.0	94.5	98.9	111.3	82.6	113.7
1981	88.8	107.3	100.0	111.9	98.6	136.0

Source: International Iron and Steel Institute (IISI), *Steel Statistical Yearbook* (Brussels, IISI: various years); Quoted in Barnett and Schorsch, 1982, p. 55.

(OTA, 1980, p. 284). Its advantages over the open hearth process center on its acceleration of the refinement of the molten charge and its shorter heat times, reducing costs and increasing productivity. The electric furnace represents another efficient method of steel production used to make certain types of steel products and will be discussed in more detail below. Table 4.3 illustrates the rate of adaptation of BOF technology and the replacement of open hearth facilities in major steelmaking countries. In order to illustrate the overall technological competitiveness of the steel industries shown, the amount of electric furnace capacity has also been added to the BOF figures in Table 4.3. While the resulting picture shows the predominance of efficient methods in steel production among major producers, the United States has continued to maintain a sizable amount of open hearth capacity, whose cost inefficiencies make it extremely vulnerable to competition from foreign and internal sources. Modern production processes are in fact used in almost all new steel capacity built since 1960, including virtually all steel plants in LDCs.

The other major advance in steelmaking technology involves the process of *continuous casting*. Traditional steel production involves the casting of raw steel into ingots, which then must be re-heated for rolling into semi-finished shapes. Continuous casting replaces these steps by pouring the molten steel directly into a mold, in which it flows

Table 4.4   *Continuous casting adoption rates 1971−83*

	US	EC	Japan	Canada	S. Korea	Brazil
In millons of short tons						
1971	5.8	6.7	11.0	1.4	—	0.1
1972	7.7	11.0	18.1	1.5	—	0.2
1973	10.2	15.5	27.2	1.7	—	0.3
1974	11.8	21.6	32.4	2.1	—	0.4
1975	10.6	22.8	35.1	1.9	0.4	0.5
1976	13.5	29.7	41.5	1.7	0.8	1.2
1977	14.7	35.3	46.1	2.4	1.5	2.2
1978	20.8	42.2	52.0	3.3	2.0	3.3
1979	23.0	47.7	64.0	3.5	2.6	4.2
1980	22.7	55.2	73.1	4.5	3.1	5.6
1981	24.5	62.4	79.1	5.3	5.3	5.3
1982	21.6	64.2	86.4	4.3	6.6	5.8
1983	25.0	66.6	86.7	NA	NA	NA
In percentage of crude steel output						
1971	4.8	4.8	11.2	11.5	—	0.8
1972	5.8	7.2	17.0	11.7	—	2.2
1973	6.8	9.4	20.7	11.6	—	3.2
1974	8.1	12.6	25.1	13.8	—	5.1
1975	9.1	16.5	31.1	13.3	19.7	5.7
1976	10.5	20.1	35.0	11.9	21.9	12.1
1977	11.8	25.4	40.8	15.9	31.7	17.4
1978	15.2	28.9	46.2	20.2	36.8	24.7
1979	16.9	30.9	52.0	19.9	30.6	27.6
1980	20.3	39.2	59.5	25.6	32.4	33.4
1981	20.3	45.1	70.7	32.2	44.3	36.6
1982	29.0	52.8	78.7	32.8	51.3	41.0
1983	29.7	55.9	81.4	NA	NA	NA

Source: Mueller, 1984, p. 123

as it cools, creating a semi-finished shape in one continuous step. Among the advantages of continuous casting over traditional methods are (1) higher yield due to the elimination of scrap waste, (2) lower energy cost due to the elimination of re-heating and (3) greater labor productivity due to the elimination of intermediate steps and handling (Barnett and Schorsch, 1983, pp. 56−7).

The cost advantages of continuous casting make it an important indicator of competitiveness on world markets. Table 4.4 shows the adoption rate of this process in the United States, Japan, the EC, Canada,

Brazil and South Korea. Among the major western steel producers, Japan has gone the farthest in its use, followed by the EC and the United States. The high rate of adoption in Japan and even in some NICs is a result of the recent vintage (late 1960s and thereafter) of the steelmaking capacity built there, since its advantages make its adoption in any new steel plant construction almost imperative (OTA, 1980, p. 288). The slow pace of adoption in the United States, in contrast, appears to be part of a vicious circle of declining competitiveness, lack of technological development, and continued competitive decline.

Cost structure
The problem of excess capacity discussed earlier comes into clearer focus with an international comparison of production costs among steel industries in competition on world markets. If steel production processes, costs and output around the world were uniform, and if transportation costs were negligible, the existence of excess capacity would mean that the long-run price was insufficient to meet the average total production costs of the typical firm. In this case, market forces would require the world industry to contract until the reduction in supply raised the long-run price to a level that would allow all remaining firms to meet total average costs. The shutdown of infra-marginal plants in this theoretical case would be distributed evenly around the world, since all plants were assumed to be identical.

Yet, in a world of technological gaps between competing steel industries, diverging production costs and differential production mixes in steel output, the problem of excess capacity implies a shake-out of the world industry that is based on relative competitive disadvantage. It makes no difference from an economic perspective who is at 'fault' for the world's excess capacity; the economic solution rests on a determination of what capacity is viable on world markets in the long run. The difficult political problem arises when existing capacity of older vintage becomes redundant in a market open to international competition, especially when foreign construction of 'excess capacity' is blamed for the depressed market. A market-driven adjustment process, however, levels no such blame and makes no such distinction between national boundaries; it is the uncompetitive capacity that must be scrapped. The private cost of adjustment in these situations provides the impetus for protecting the cost disadvantage through trade restrictions.

Labor cost, technological diffusion and construction cost The differential rates of adoption of key technological innovations, combined with the international structure of labor and other input

Table 4.5   *An international comparison of employment costs (blue and white collar employees, including contract workers) (1960 and 82)*

	Hourly Employment Costs		Hours per Net ton of Steel Shipped		Unit Labor Cost	
	1960	1982	1960	1982	1960	1982
USA	3.95	22.74	17	10.3	67	234
EC	1.16	11.20	21	10.1	24	113
Japan	0.54	10.18	51	8.3	28	85
Brazil	0.60	4.00	NA	20.0	NA	80
South Korea	NA	2.65	NA	14.0	NA	37

Sources: Bureau of Labor Statistics, unpublished data; AISI, *Annual Statistical Report*, various issues and *Steel Employment News*, January 11, 1983 (white-collar employee differential estimated); Eurostat, *Wages and Incomes*, various issues; Ministry of Labor, Japan, *Monthly Survey of Labor Statistics*; NRI (a consulting firm) Tokyo; Tekko Roren, Rodo Handbook, 1981, and information obtained from the Instituto Brasileiro de Siderurgia and the Grupo Industrial Alfa, Mexico. Coking Coal and Iron Ore rates and costs were estimated from the standard industry sources (AISI, *Annual Statistical Report*, IISI, *Statistical Yearbook Steel*, JISF, *Tokei Tekko Yoran*, and *Siderurgia Latinoamericana* and from *Metal Bulletin*, various issues); rates per ton of hot metal were then transformed by the use of hot metal/raw steel and raw steel/finished steel ratios estimated from IISI, World Steel in Figures, 1981, and German Steel Federation, *Statistisches Jarbuch*, 1983; quoted in Mueller, 1982, p. 120.

costs, have produced a changing pattern of competitiveness on world steel markets. Hourly wage rates for steel workers, for example, have been consistently highest in the United States for many years, followed by Europe and Japan among major steel producers, and have remained several times the wage rates in the NICs. This factor did not pose a major problem for the high-wage countries as long as labor productivity, based on superior skills and technology, offset the wage differential. The diffusion of advanced steelmaking technologies to lower wage countries, however, as well as the sluggish adoption of new technologies in the higher-wage markets, have created a substantial labor cost gap. Crandall (1981, pp. 78–81) has also noted the relatively low cost of constructing new steelmaking capacity in Japan and eastern Asia in general; this factor undoubtedly allowed steelmakers in this region to take advantage of new technologies. Table 4.5 compares hourly wage rates, labor hours per net ton of steel shipped and unit labor costs between the United States, the EC, Japan, Brazil and South Korea in 1960 and 1982. The US/Japan and EC/Japan gap in unit labor costs has risen over the period, one indication on Japan's increasing competitive advantage. Yet the absolute advantage in unit labor

cost now resides in the NICs, whose modern steelmaking facilities and low wages have lowered their labor cost below those in Japan.

In addition to absolute differences in unit labor costs among countries, a key to the impact of labor costs on steelmaking competitiveness lies in the trend of productivity and wage rate increases over the years. In most countries, compensation to steelworkers exceeds the manufacturing average. The steel wage premium in 1982 averaged about 20 per cent in Europe, 25 per cent in Canada and 60 per cent in Japan (CBO, 1984, p. 23). In the United States, the premium was less than 50 per cent until the early 1970s, when a new collective bargaining framework between the American steel industry and the United Steelworkers (USW) Union, called the Experimental Negotiating Agreement (ENA), was adopted. The ENA provided for guaranteed wage and cost-of-living increases in exchange for a no-strike agreement. Because of the bargaining power of the union and the oligopolistic structure of the American steel market, labor cost increases could be passed on to consumers, a practice that was later made impossible by international competition. The ENA appeared to contribute to a growing gap between productivity and wage rate increases, as illustrated by Table 4.6. The wage-productivity gap in the US steel industry became particularly damaging during the period from 1972 to 1977, when the increase in hourly earnings (68 per cent) far exceeded productivity gains (3 per cent), a gap much larger than that for US manufacturing as a whole. A later study, based on 1982 statistics, estimated that US steel wages were so severely excessive that they would have to fall 40 per cent in order to match labor costs in the Japanese steel industry (Kreinin, 1985, p. 184). Not only did this exacerbate the deterioration of American steelmaking competitiveness, but also caused the USW to become an entrenched protectionist interest group, whose natural political interest lay in consolidating the gains it had won. While the ENA was abandoned in 1982 and some reductions in USW compensation have occurred, premium wage rates remain a major source of competitive disadvantage for the American steel industry.

Other input and transportation costs An important source of comparative advantage for the established steel industries of the United States, and to a lesser extent, Europe, used to be their proximity to iron ore and coal deposits, important inputs in the production of steel. Integrated American steel producers, in particular, owned iron ore and coal mines in the Great Lakes region of the United States, thus assuring themselves of captive supply at low cost.

However, several factors have eroded the cost advantages of these producers over the years. Many of the newer producers, for example,

Table 4.6 *Percentage increase of average hourly earnings (current dollars) and in output per hour of labor input, selected periods*

	Hourly earnings		Output per hour	
	All workers [a]	Production workers	All workers [a]	Production workers
All manufactures [b]				
1957–77	195	182	69	NA
1957–67	43	40	33	NA
1967–72	35	35	16	NA
1972–77	53	49	9	NA
Steel and steel products [c]				
1957–77	224	227	37	47
1957–67	36	34	19	23
1967–72	42	43	13	14
1972–77	68	70	3	5

Source: Anderson and Kreinin, 1981, p. 202. Calculations by the authors from data in *United States Census of Manufactures* for 1957, 1967, 1972 and 1977, Bureau of the Census, United States Department of Commerce, Washington, for hourly earnings; *Handbook of Labor Statistics*, Bureau of Labor Statistics, US Department of Labor, Washington, 1978, for output per hour in aggregate manufacturing; and *Productivity Indexes for Selected Industries*, Bureau of Labor Statistics, US Department of Labor, Washington, 1979, for SIC 331 and 371.

[a] Non-production workers are assumed to work the same annual hours as production workers.

[b] Output originates from gross domestic product (GDP).

[c] Standard Industrial Classification (SIC) 331; 'output' is a physical production series constructed by the Bureau of Labor Statistics.

have built their steel plants in close proximity to deepwater ports, providing for easy access to ocean shipping facilities and foreign markets. The cost of transoceanic shipping has declined substantially since the 1950s, making it easier and less costly for these steel producers to receive imported raw material inputs, such as iron ore, and to export steel products to foreign markets. This factor was especially important for countries such as Japan, which were dependent on foreign sources for many of the raw materials needed for steel production and which required access to foreign markets to expand production and achieve economies of scale. The locational advantage was enhanced by the completion of the St Lawrence Seaway, which provided foreign steel producers with access to the large American interior market (Crandall, 1981, p. 23).

Trends in input sourcing, combined with the transportation factor, have also led to competitive shifts. Deposits of high-grade iron ore

have been largely depleted in Europe and the United States. New sources of iron ore have been discovered in Venezuela, Brazil, Australia, Canada and Africa (Hogan, 1983, p. 8), but then the cost of shipping and the location of steel facilities became more important in determining the *delivered* cost of inputs. Under these circumstances, Japan and some NICs with steel plants close to deepwater ports have benefited from relatively easy access to foreign supplies of high-grade ore, while the United States has come to rely on lower-grade Canadian ore and inland EC producers have suffered from increased transportation costs of imported ore. Another factor in the input cost equation is the differential efficiency with which iron ore and coking coal are used. Newer and more efficient steel plants can offset the higher price of the input in the final unit cost. Finally, the cost of energy in the world's steelmaking areas since the OPEC oil price shock of 1973 has tended to converge, even though the United States has retained an advantage in this category, based largely on its supplies of coal (CBO, 1984, pp. 24–5).

Table 4.7 tabulates coking coal and iron ore prices and unit costs for steelmakers in the United States, the EC and Japan from 1960 to 1981. In coking coal, the United States has maintained a delivered price advantage, but this is wiped out in final unit cost, where Japan has gained a slight advantage. Iron ore prices for US producers, on the other hand, which were once lowest in the world, became higher than those of its major international rivals by 1970, and final unit costs have risen accordingly.

Table 4.8 compares the major input costs of the United States, the EC, Japan, Brazil and South Korea as they stood in 1982. While the large appreciation of the US dollar since early 1982 has contributed to the gap between the cost figures for US producers and their international rivals, the US steel industry on average remains at a severe cost disadvantage on world markets. EC production costs, in turn, have persistently remained above those of Japan, which remains the lowest-cost producer among major steelmaking countries. South Korea, however, with its modern, efficient steel industry, has become the lowest-cost producer overall, and has even begun to penetrate Japanese markets.

Entrepreneurial lassitude
The signs of competitive decline among integrated firms is in many cases a reflection of poor planning and decision-making on the part of the firm's managers. This issue has been the focus of much study in the United States in particular, where there was a lack of a vigorous competitive environment in the steel industry for many years. As early as the 1930s, a management consulting firm investigating the perform-

Table 4.7 Unit costs of major inputs 1960–81 ($ per net product ton)

	Employment costs			Coking coal			Iron Ore			Total of three input costs		
	US	EC	Japan	US	EC	Japan	US	EC	Japan	US	EC	Japan
1960	62	23	27	12	14	11	13	14	16	87	51	54
1965	58	29	24	9	14	10	13	11	18	80	54	52
1970	74	33	22	14	14	13	16	10	16	104	57	51
1975	124	85	51	34	38	38	34	22	26	192	145	115
1978	148	130	76	38	40	39	38	30	31	224	200	146
1981	190	135	85	63	62	57	50	42	45	303	240	187

Sources: For employment and raw materials costs 1960–1975: Federal Trade Commission, Steel Study, 1977, pp. 113–18; Council on Wage and Price Stability, Report to the President on Prices and Costs in the United States Steel Industry, Washington, DC, October 1977, pp. 126, 127, 139, 140; Mueller, H. and Kawahito, K., 1978, Steel Industry Economics: A Comparative Analysis of Structure, Conduct and Performance (New York): Japan Steel Information Center), pp. 18–20.

For employment and raw materials costs 1980: trade publications, steel federation statistics, Eurostat, Iron and Steel series; Charles Bradford, Steel Industry, Quarterly Review (May 1980); Marcus, P. and Kirsis, K., 1979, World Steel Dynamics, Core Report J (New York: Paine Webber Mitchell Hutchins), Tables 3 and 5.

Quoted in Adams and Mueller, 1982, p. 121.

Table 4.8   *An international comparison of major input costs 1982*

	Labor	Coking coal and iron ore	Energy	Total	Difference from USA cost
USA	234	103	72	409	zero
EC	113	100	62	275	134
Japan	85	90	64	239	170
Brazil	80	95	65	240	169
South Korea	37	90	66	193	216

Source: see Table 4.5.

ance of US Steel Corporation had concluded that the company had 'inadequate knowledge of its domestic markets and no clear appreciation of its opportunities in foreign markets ... less efficient production facilities than its rivals ... [and was] slow in introducing new processes and new products' (Stocking, 1950, p. 957, quoted in Adams and Mueller, 1982, pp. 116–17). More recently, American integrated steelmakers were unprepared for the shifting pattern of steel consumption in the late 1970s that resulted in a surge in demand for certain types of steel, especially pipe and tubular goods, while demand in other product lines declined (see Mueller, 1984, p. 130). The evidence furthermore suggests that integrated US steel firms as a whole have been followers rather than leaders in research and development and process innovations (Adams and Mueller, 1982, pp. 111–19). The stock market performance of steel equities has apparently reflected decreasing confidence in the industry since competitive decline became evident in the 1950s (Crandall, 1981, pp. 28–30).

Much of the debate over industry performance has focused on the use of new technologies. Some studies indicate, for example, that American steel producers were sluggish in adopting BOF technology (Adams and Dirlam, 1966; Baumann, 1974) and continuous casting (Ault, 1973), largely because there was initially no competitive pressure to do so. The steel industry and its supporters have claimed, however, that early adoption of BOF technology was not economically feasible at the time because of the new installation of open hearth furnaces in the 1950s (see FTC, 1977, pp. 483–8) and that adoption was later hindered by a 'lack of capital' due to chronically depressed prices. Yet from an economic point of view, the economic value of the new open hearth furnaces was greatly diminished by the development of BOF technology. Although it would have been painful to write off a large portion of this investment as a loss, a market-driven response would have dictated it. As increased competition, especially from

abroad, began to threaten the industry, only import controls could protect the book value of the obsolete equipment. The alleged 'lack of capital' also provides a weak explanation of the industry's poor performance. In point of fact the industry did engage in expensive investment programs in plant and equipment in the late 1960s and late 1970s, indeed outspending its European and Japanese rivals in this area. None the less, such investments 'foundered on the lack of integrated, organic company planning with respect to plant structure and organisation — a deficiency that continues to plague the majority of integrated steel companies [in the United States], (Adams and Mueller, 1982, p. 119).

Other sources of questionable business strategy among US integrated producers center on the inattention paid to cost competitiveness, a common trait of a traditionally oligopolized industry (see Barnett and Schorsch, 1983, pp. 60−74). The oligopolistic structure of the American steel industry, marked by patterns of price leadership, has been well documented (FTC, 1977, pp. 152−70 and citations therein). One striking feature of this system has been its ability to maintain price discipline in times of depressed demand. The structure of the domestic market, insulated from trade disturbances until the 1960s, not only gave firms the ability to consistently maintain steel prices above marginal cost but also allowed them to internalize such pricing behavior as normal and 'fair.' As a result, steel capacity was kept in operation in excess of what would have been possible under more competitive market conditions. The viability of this previously insulated market structure and the associated profitability of marginal (mostly ageing) steelmaking facilities was finally challenged by the rigorous standards of an increasingly competitive world steel market. Years of isolation from the world market made United States producers reluctant to adjust to more stringent competitive conditions, particularly to the anathema of price-cutting. Thus, the American industry was all the more vulnerable to a cyclical trough in steel demand when its approach to recession pricing strategy clashed with the world export market's response.

As long as competition from abroad (and at home) did not impinge on domestic markets, price increases could be passed along to consumers. Instead of an economic strategy to reduce costs, the steel industry has followed a political strategy to reduce import competition that has required an alliance with labor and, along with it, the persistence of the labor cost disadvantage. Two former economists at the American Iron and Steel Institute, the industry's trade organization, have summarized the cause of the industry's decline by stating that they have failed to grasp the absolute imperative of dynamic cost competitiveness, maintaining increasingly disadvantageous sources of raw materials and

trading wage restraint for the sake of a protectionist cabal in Washington' (Barnett and Schorsch, 1983, p. 74).

In contrast, market structure and competitive pressures in Japan and the EC have forced cost consciousness on most steel producers there. In Japan, the steel industry is in fact oligopolistic, but pricing discipline has been elusive. Even in the absence of import competition, Japanese firms suffered losses on home market sales between 1975 and 1977, indicating that prices could not be maintained in times of declining domestic demand (Adams and Mueller, 1982, p. 106). In addition, the export orientation of the Japanese steel industry, which competes on the world market for customers, forces cost efficiency measures and competitive pricing practices among Japanese producers. Since the founding of the European Coal and Steel Community in 1952, when trade barriers between the member countries began to decline, the European steel industry has become increasingly competitive, and several episodes of price-cutting occurred during the 1960s and 1970s until minimum price rules were imposed (see Stegemann, 1977, Chapter 3). This situation naturally causes firms to concentrate on cost competitiveness. In the postwar period, the European industry has been too dispersed and multifarious to permit any co-ordinated pricing discipline, a characteristic that has also made operation of the Community-wide steel cartel, Eurofer, difficult.

Among steel producers in the EC, signs of entrepreneurial lassitude appear to be a more recent phenomenon linked with increasing government involvement in the industry. Research has linked increases in government ownership with decreases in technological adaptation (FTC, 1977, Chapter 7). Furthermore, government policies may prevent otherwise market-driven decisions to cut capacity and lay off workers, a problem the British Steel Corporation apparently faced, for example (FTC, 1977, p. 503).

Perhaps the most serious threat to the vitality of the traditionally competitive steel industry of the EC, however, has been the systematic elimination of price competition under the cartel arrangements of the European Commission's manifest crisis plan. The policies of the Commission enforce a 'burden sharing' of production cuts among steel producers and have attempted to maintain minimum prices. The inevitable result is that more efficient firms must yield market shares to less efficient firms, a situation that effectively validates the maintenance of uncompetitive capacity. Small, competitive steel producers in the Bresciani region of Italy, for example, were fined for violating minimum price regulations, despite their plea that the crisis plan discriminated against efficient producers. The European Court of Justice upheld the fines, declaring that such considerations were outweighed by the collective well-being of the industry (Trainer, 1980, pp.

605–6). The rigid institutional framework of cartels removes much of the incentive structure that promotes vigorous adjustment to competitive change. In this manner, the retirement of redundant steel capacity in the EC has been delayed, sustaining the vulnerability of the industry to international competition.

• *The emergence of mini-mill producers*
The analysis of the chapter thus far has treated the steel industry as a homogeneous collection of firms with similar production processes. Since the 1960s, however, a new group of producers, distinct from the large, integrated companies, has emerged: mini-mills. Integrated producers generally combine the discrete steps of steelmaking on a large scale, involving high capital costs, to make a wide range of steel mill products. Mini-mills, in contrast, operate on a smaller scale, using electric arc furnaces, continuous casters, and a rolling mill to produce a limited range of steel products (Barnett and Schorsch, 1983, pp. 83–5). The mini-mills have been an important source of competitive change in steel markets, due to their flexibility, advantageous location and lower costs. The cost structure of electric arc furnace production has made the United States the area of greatest mini-mill expansion, since steel scrap and electricity, the main inputs, are available there most cheaply. Mini-mills also form a significant or growing part of the steel market in Japan, Italy, West Germany, the Netherlands, the United Kingdom, Canada and Australia.

At present, mini-mill technology limits production to a small number of products, principally wire rod, structural shapes and bar products, although new technologies may make a wider range of production possible in the future (OTA, 1980, p. 253). In these particular products, however, they have captured large portions of the market from integrated producers. While mini-mills accounted for about 18 per cent of US steel production in 1982, their market share was approximately 95 per cent in bar-size light shapes, 75 per cent in concrete reinforcing bars and 53 per cent in wire rod (Barnett and Schorsch, 1983, p. 88). The main source of their competitive advantage lies in lower production costs. Table 4.9 compares integrated and mini-mill per-ton costs of wire rod production. In every category of inputs – raw materials, energy, capital and labor – mini-mill costs are lower than those of integrated mills (see Barnett and Schorsch, 1983, pp. 94–5). In the United States, labor cost savings in mini-mill production come from both higher labor productivity and a largely non-unionized workforce. Another source of competitive advantage lies in the advantageous location of the mills in close proximity to their consumer markets. Many mini-mills are located in the southern, southwestern and western areas of the United States, locations not

Table 4.9 *Comparative production costs in 1981* [a] *wire rods (dollars per net ton shipped, normal operating rates* [b]

| | Integrated | | | Mini-Mill | | |
	USA	West Germany	Japan	USA	West Germany	Japan
Labor	131	84	51	60	45	37
Iron ore	62	50	49	—	—	—
Purchased scrap [c]	15	5	3	93	96	96
Coal/coke	52	75	59	—	—	—
Other energy	46	37	40	45	52	51
Other costs [d]	60	61	64	65	69	68
Oper. costs	372	312	266	263	262	252
Depreciation	12	14	16	11	12	11
Interest	5	8	18	7	8	10
Misc. taxes	5	2	4	3	1	2
Total costs	393	336	304	284	283	275 [e]

[a] Exchange rates in 1981, at 2.26 DM/$ and 230 Y/$, were somewhat out of line with historic relationships. As a result, US production costs, especially relative to those estimated for Europe, may be slightly overstated.

[b] For integrated plants, average 1977 to 1981 capacity utilization rates were used to avoid single-year abnormalities. For the United States, capacity utilization averaged 80 per cent, for West Germany and Japan, 65 per cent. For mini-mills, 85 per cent capacity utilization was assumed throughout, and this is close to their average over the last five years.

[c] Average 1980–81 scrap prices were used, since these are more typical of long-term relationships than 1981 scrap prices alone.

[d] Includes alloying agents, fluxes, refractories, rolls and so on.

[e] Excluding any return on equity.

Sources: Estimated by Barnett and Schorsch, 1982, p. 95, from data contained in annual reports (e.g. Korfstahl, Tokyo Steel, Florida Steel, etc.), Metal Bulletin, World Steel Dynamics, *Core Report Q*, 1982 (New York): Paine Webber Mitchell Hutchins, and so on.

conveniently served by integrated producers (Mueller, 1982, p. 25).

An examination of mini-mills puts the issue of steel industry adjustment into clearer focus. In the United States it can be said that there are two steel industries: an integrated sector that has adapted poorly to changing competitive conditions and a mini-mill sector that has adapted well. Without the burden of an oligopolistic heritage, mini-mills have focused on the importance of cost competitiveness and technological change, acting as price-takers faced with the constant threat of competition. Aside from outperforming the integrated producers in terms of profitability (CBO, 1984, p. 31), they have done well

in competition with imports that have otherwise made extensive inroads into US steel markets. Their financial performance suggests, furthermore, that future investment funds will be available to allow continual technological adaptation. In short, mini-mills have not only been a *domestic* source of competitive decline for integrated producers; more significantly, their market performance has also provided the example of an alternative to protectionism as a response to competitive change in steel markets.

The question of government subsidies to steel

The controversy over the impact of government subsidies to national steel industries on the pattern of steel trade and steel prices has become a major issue in trade relations in recent years. The American steel industry based its call for protection in 1984 on allegations of massive government subsidization among its foreign competition and has put considerable effort into documenting such practices (AISI, 1984). This is a difficult and complex issue for which no definitive answers are yet available, largely because of the lack of up-to-date economic research on the subject. This study can only attempt to provide a framework for discussing the issue and offer a summary of the available evidence.

There are both economic and political issues surrounding the subsidies controversy. First and foremost among the economic issues is how much government intervention actually affects steel output, trade and prices. Even if subsidization exists, economic analysis remains skeptical of the damage that it does to the importing country. As was shown in Chapter 1, subsidized imports must be damaging to the economy as a whole (not just to import-competing producers) in order to justify trade restrictions on economic grounds. Political considerations override the economic questions, however. National trade laws and the GATT allow the use of countervailing duties against subsidized exports as long as it can be shown that domestic producers are suffering or are threatened with injury as a result (the Subsidies Code concluded at the Tokyo Round waives the injury test for exports from non-signatories to the Code). Aside from these legal (as opposed to economic) criteria for trade restrictions, subsidies are of serious concern in terms of the damage they can do to trade relations in general.

The nature of government involvement
Government involvement in national steel industries has increased substantially in recent years. Government ownership of steel mills has generally increased since 1974 (with some exceptions, notably the

Table 4.10   *World raw steel production, total and by major state-owned firms, 1982 (millions of net tons)*

	Total production	Production by state-owned firms [a]	State-owned percentage of total
USA	75	0	0.0
Japan	110	0	0.0
FRG	40	7	16.4
Italy	26	15	55.4
France	20	17	84.2
United Kingdom	15	13	83.2
Belgium	11	8	72.7
Netherlands	5	0	0.0
Luxembourg	4	4	100.0
Total [b]	121	63	51.7
Other Western developed [c] of which:	60	25	42.0
Canada	13	0	0.0
Spain	14	5	34.4
South Africa	9	7	78.8
Western developing Total of which:	61	35	57.4
Brazil	14	8	54.6
Mexico	8	4 [d]	50.0
Venezuela	3	2	87.0
South Korea	13	10	74.6
India	12	7	60.9
Total Western world	426	122	28.6
Non-market economies	271	271	100.0
All other [e]	14	0 [f]	0.0
Total	710	393	55.4

Note: Totals may not add due to rounding.
Sources: IISI, *World Steel in Figures, 1982*; Metal Bulletin, *Iron and Steel Works of the World*, 8th edition, 1983; Compiled in: Department of Commerce, 1984.

United Kingdom), and many government-run mills from NICs have begun to play an important role in world steel trade (Table 4.10). Furthermore, governments have often treated national steel industries as a means for achieving national goals and social policies. In order to assure military and economic well-being, for example, governments have promoted the expansion of the steel industry, either through admonishments or financing incentives. Employment in the steel industry is in many countries a matter of national social policy, where direct or indirect controls hinder companies from laying off redundant workers. Government controls have extended to steel pricing policies. Typically, government 'jawboning' or direct controls are applied if prices rise to 'unacceptable' levels, while government subsidies or import controls (or both) are used to help prop up weak prices (Mueller, 1982, p. 3).

The effect of government involvement in the steel industry has been to create an atmosphere of protectionism, based not only on the perceived importance of the domestic industry, but also on the perceived advantages that foreign rival steel industries receive from their governments. There is, in the current vernacular, no 'level playing field' in steel trade, since allegedly subsidized firms can presumably expand exports and capture foreign market shares at the expense of domestic firms. Yet whatever role governments play in determining the level of steel capacity, output, employment and pricing, the question of trade-distorting government subsidies requires a careful economic determination of the effect of such programs on the *cost* of inputs. Economic theory predicts, in other words, that only a distortion in the unit economic costs of steel production or restrictions placed on domestic deliveries will distort steel trade patterns. Government equity in the steel industry, in and of itself, may not necessarily alter steel mill operating costs; government exhortations and policies may or may not have an impact on the market. A reliable subsidy investigation

---

[a] Data was available only for the top 48 (based on raw steel production) Western firms.

[b] Total only for those EC countries listed. Data for Denmark, Ireland and Greece are included in 'all other.'

[c] Canada, Australia, South Africa, Spain, Austria, Sweden, Finland and Yugoslavia.

[d] USDOC estimate.

[e] Countries which produced less than 2 million tons in 1982.

[f] In the absence of complete data, we assume zero government ownership. As a result of USDOC CVD investigations, we know that some small firms such as ISCOTT, the steel company of Trinidad and Tobago, are 100 per cent government owned. Even if all firms in this category were 100 per cent government owned, including those in Denmark, Ireland and Greece, the figure for total world state-owned production would increase by only 2 per cent.

must therefore not only identify government involvement, but also quantify its net effects in altering market-driven outcomes (FTC, 1977, pp. 310–16).

The difficulty of this task becomes clearer upon consideration of the possibilities of government intervention in a complex economic system. Clear-cut cases, such as direct subsidies on exports, are rare, since these are directly countervailable according to GATT article VI. Yet the possibilities of government intervention altering market outcomes are nearly endless. A representative list would include government programs that provide: production subsidies, grants and compensation for operating losses, loans at reduced interest rates, loan guarantees, buy-domestic rules, foreign aid tied to domestic purchases, regional development grants, worker training programs and tax relief and incentives (Adams and Mueller, 1982, p. 126). In addition, one must consider any government efforts to provide or develop infrastructure, such as waterways, communications and roads, that are of benefit to the industry. Any government policies that allow domestic producers to receive higher prices and/or increase output, including trade restrictions, also act as subsidies. In the latter case, the subsidy is financed directly by the domestic consumers of the product.

Even if one compiles a satisfactory list of public assistance activities, however, assigning values to them is often difficult, if not impossible. In the steel industry, in particular, government 'aids' may be illusory. The American steel industry, for instance, has alleged that foreign governments, particularly Japan, targeted the steel industry for growth, funneling investment funds towards the construction of steel, and directing the development of the industry. In the case of Japan, the steel industry actually received only a small amount of direct government subsidies; the main source of investment appeared to come from the private sector, but only after government planners had 'designated' steel as a growth industry (see Johnson, 1982, p. 211; Sakoh, 1983). How can one assign subsidy values to such government activity? In general, the mistake is often made of confusing institutional arrangements for subsidies (Kawahito, 1981, p. 240). The problem lies in trying to determine what would have happened if the government had done nothing. An industry that possesses the tools for competitive success on world markets would presumably grow and proper without the government's seal of approval.' Theoretically, the impact of such government policies must therefore be at least partially discounted.

In order to calculate net subsidies, it is also necessary to consider the 'proportionality effect' and the impact of offsetting negative subsidies. Proportionality refers to the fact that across-the-board subsidies or other government intervention affecting *all* industries will not affect

the competitiveness of a particular industry. This is because exchange rate and general price level movements tend to offset the general effect of broad-based subsidies (FTC, 1977, p. 314). Negative subsidies refer to government policies that actually *increase* the costs of affected firms. In the steel industry, government policies may prevent or restrict the ability of firms to lay off workers, or may require them to use domestic sources of inputs. One must subtract the impact of negative subsidies from that of positive subsidies to calculate the effective net rate of subsidization (see FTC, 1977, Chapter 6).

In 1977 the US Federal Trade Commission attempted a comprehensive study of subsidies in national steel industries, covering government programs for most countries through 1975. The results implied that, for the period covered, government intervention had at most a minimal impact on international competitiveness in the steel industry. For the United States and Belgium, estimates of the net subsidy value per ton of steel were zero. Using an average price of $275 per ton, the subsidy rate per ton for Japan and France was 0.5 per cent or less. Germany's estimated net subsidy rate was actually negative, −1.4 per cent, based on the strong influence of the West German government's requirement that domestic steel firms use higher priced domestic coal. In only two countries were subsidy rates estimated at one per cent or more: Italy (1.0 per cent) and the United Kingdom (4.5 per cent) (FTC, 1977, p. 369).

The deepening recession in the steel industry since the publication of the FTC study has, however, made its estimates obsolete, at least for some countries. The Belgian government acquired 50 per cent ownership of the two largest national firms in 1978. In France, the two largest steel companies were effectively nationalized in 1979. Government grants covering operating losses increased the level of subsidization in these countries. Further subsidies in the form of trade restrictions have been granted in the United States (the trigger price mechanism of 1978−82 and the steel pacts of 1982 and 1984) and in the EC (the basic price system and associated VERs since 1978). In addition, many NICs, whose steel industries are largely government owned, have increased production and exports since the mid-1970s. Little research exists, however, on the impact of government involvement on the overall international competitiveness of steel industries in the NICs.

Recent decisions of the US Commerce Department on countervailing duty cases provide another view of the subsidies issue, although the economic validity of the criteria used is questionable. This is because the investigations do not allow for offsetting negative subsidies or proportionality effects, which must be considered in order to calculate *net* subsidy margins (see Mueller and van der Ven, 1982, pp. 265−6). Commerce Department officials appear, in fact, to be aware of these

shortcomings, but there is no internal mechanism to change the standards and criteria used in the investigations. In this regard, it is important to place the role of countervailing duty laws and associated bureaucratic guidelines in their proper political perspective. As was discussed in Chapter 1, these and other 'unfair' trade laws are designed to provide import-competing industries with legal channels for seeking trade restrictions based on trade practices deemed unacceptable by the GATT. They are in large part a sop for protectionist sentiment that allows governments to pursue trade policy objectives unencumbered with constant petitions for protection based on charges of 'unfair' trade. This characteristic may help to explain why the laws do not in general reflect economic criteria. From the government's point of view, a stricter adherence to economic standards may be undesirable for the very fact that such an approach would make protection from such legal trade barriers more difficult to achieve, a phenomenon that would only tend to re-channel protectionist sentiment towards more direct means of trade restriction, such as legislative tariffs and quotas. In this manner, the countervailing duty laws tend to be biased towards domestic producer interests, and one must view the calculated subsidy margins with this caveat in mind.

Table 4.11 tabulates recent countervailing duty determinations by the Commerce Department and the final subsidy rates calculated in each case. The most significant cases include those filed against EC steel industries in 1982, since these covered the largest volume of steel trade and gained the most political attention. The subsidy rates are larger than those calculated by the FTC, the result in part of increased government activity in the industries under investigation since 1975, in part of the narrower product coverage used and the methodology biased towards higher margins discussed above. Still, the Commerce Department investigation found almost no countervailable imports from the largest EC supplier, West Germany, and found very small or *de minimis* subsidy rates on imports from the Netherlands and Luxembourg.

Even in cases where the evidence has pointed to the existence of substantial subsidies using government criteria, however, the impact of the subsidies themselves on the plight of the steel industry has probably been very small. Commissioner Stern of the ITC, who in fact voted in favor of import restrictions covering most steel trade involved in the 1982 countervailing duty cases, estimated none the less that even under assumptions most favourable to US steel producers, countervailing duties would only have 'saved' 2,259 jobs, less than 1.5 per cent of unemployed steelworkers at the time (ITC, 1982b, pp. 69*–70*). She concluded that 'the overall problems of the steel industry have very little to do with the subsidized European imports under investigation' (ITC, 1982b, p. 70*).

It is therefore clear that the subsidies issue has a much greater

Table 4.11 *Recent affirmative subsidy determinations (final), US Department of Commerce*

Country	Products	Subsidy margin (%)	Date
Argentina	2	2.34–6.42	18 April 1984
Belgium	18, 10, 13	0.348–13.411	7 September 1982
Brazil	16	12.53	24 August 1982
	14	13.90	15 October 1982
	11, 12, 3	17.49–62.18	18 April 1984
France	14	4.792	15 October 1982
	13, 3, 5, 18	3.702–24.416	7 September 1982
West Germany	18	1.131	7 September 1982
	10, 13, 3, 5	*de minimus*	7 September 1982
Italy	3, 5, 13	6.32–14.56	7 September 1982
Korea	10, 12, 6, 19	0–1.88	20 December 1982
Luxembourg	18	0.539–1.523	7 September 1982
South Africa	14	27.1	17 May 1982
	18, 10, 12, 5	6.7–15.1	23 August 1982
	20	7.8/*de minimus*	21 September 1982
	17	10.05/*de minimus*	25 October 1982
	15	21.64	20 May 1983
Spain	14	1.77	28 June 1982
	18, 10, 3, 6		
	9, 1	0–38.25	8 November 1982
	20	16.03–29.94	8 May 1984
Trinidad and Tobago	20	6.738	27 December 1983
United Kingdom	18, 10, 8	1.88–20.33	7 September 1982

Key: (1) cold-formed carbon bars; (2) cold-rolled flat-rolled products; (3) cold-rolled sheet; (4) cold-rolled sheet strip; (5) cold-rolled strip; (6) galvanised sheet; (7) galvanised wire strand; (8) hot-rolled bar; (9) hot-rolled carbon bars; (10) hot-rolled plate; (11) hot-rolled plate in coil; (12) hot-rolled sheet; (13) hot-rolled sheet and strip; (14) PC strand; (15) pipe and tube; (16) plate; (17) rebars; (18) structurals; (19) welded pipe and tube; (20) wire rod; (21) wire rope.
Source: Department of Commerce, 1984; *Federal Register.*

impact on trade *policy* than it does on trade itself. From 1980 through 1984, approximately 70 per cent of steel imports to the United States came from countries with little or no public ownership of the steel

industry, in which little or no evidence of subsidization existed: Japan, Canada, West Germany, the Netherlands and Luxembourg. Yet, as the analysis in Chapter 7 will show, the 1982 subsidies investigation, as well as other trade law investigations in 1984, led directly to comprehensive steel trade restrictions on exports from all of these countries (and others) to the United States. The steel 'pacts' covered the entire range of steel products, even though determinations of injury or subsidization covered only limited sub-sectors of the industry; in the case of the EC pact, even EC member countries cleared of the subsidization charges were forced by political factors to take part. In general, countries are often willing to submit to export restraint in exchange for a termination of countervailing duty or other trade law investigations because of the possibility of further administrative or legislative protectionist actions that might be directed against them in the absence of a negotiated agreement.

The argument that subsidization itself has drastically altered steel trade patterns is therefore weak. Economic analysis, in any event, remains skeptical of the ability of public treasuries to significantly alter market outcomes and cause significant injury to import-competing firms over time, since the subsidies would have to be massive, long-lasting, and affect large amounts of steel exports. Subsidization, at least in the form of direct public assistance to steel industries, also appears to have a limited political life; the continual drain of public funds into steel industries in some countries of the EC has led to an official policy there (and increasing political pressure) to end operating subsidies by the end of 1985.

Whatever the result of efforts to reduce government intervention, the subsidies issue promises to remain a part of steel trade policy well into the future. As the adjustment crisis continues in the steel industries of advanced industrialized countries, their governments will continue to view any *foreign* government involvement in the steel industry with suspicion. The issue is of particular importance in the case of the NICs, many of which may have developed competitive steel industries but whose government involvement exposes them to charges of unfair trade. Only a consistent and internationally acceptable application of trade laws can create a framework of rules to bring the protectionist use of the subsidies issue under control.

Summary

The world steel industry has experienced a series of rapid and dramatic changes in the postwar period, creating competitive shocks that have challenged established steel producers to adjust. The main sources of

pick minimills And get Buck

competitive change have been (1) the expanding and increasingly dispersed nature of world steel production and exports, (2) a secular decline in steel demand, (3) technological advances in steel production, (4) the changing international structure of input and other costs, (5) public and private mismanagement in many steel firms in industrialized countries and (6) the emergence of mini-mills. While public subsidies are alleged to be a major factor in distorting steel trade, the available evidence suggests that they play at most a minimal role in international steel competitiveness, although more research is needed on this issue. In the short term, international competitive advantage in steelmaking will probably remain with Japan and some of the NICs; in coming years, the advantage of the NICs is likely to grow.

It is noteworthy that international competition has represented more a symptom than a cause of the adjustment crisis among established integrated steel producers in the United States and the European Community. The sources of competitive change in world steel markets in the postwar period have indeed been multifarious; purely 'international' factors, such as the general expansion of world steel-making capacity and a shift in cost factors favoring producers such as Japan and some NICs, cannot be blamed for these steel firms' plight. Nor has the level of government ownership in these and other countries been a decisive factor. It was rather a failure to adapt to the totality of changing market conditions that lay at the root of the adjustment crisis. Each source of market change indicated the need for a market-driven response. In the case of the established steel industries of advanced industrialized countries beginning in the 1960s, the indicated channels of market-driven adjustment included: timely technological adaptation, the retirement of obsolete or otherwise uncompetitive capacity, aggressive cost-cutting measures, increased specialization in a more limited range of steel products (such as specialty steels) and improved management performance. In general, the greatest shortcoming of steel firms in distress was the lack of a broad, international outlook on the marketplace. For several years, many established steel producers enjoyed unchallenged market positions due to technological superiority, favorable demand trends, natural trade barriers and a lack of vigorous foreign competition. When these conditions changed, these firms were unprepared to meet the challenge of adjustment.

For many steel firms, however, the potential cost and sacrifice of economically determined adjustment was high, since it would have required a severe decline in market share and firm size, possible re-organization of the steel industry and replacement of existing steel management, and in some cases, bankruptcy. The reluctance of many steel firms to pursue market adjustment channels motivated them to pursue political channels to avoid adjustment. The concerns of steel

labor unions and of governments over the implications of competitive decline reinforced the industry's case for protection. The steel industry's response to competitive shock thereby moved into the political arena, where the debate over protectionism, subsidization and cartelization began.

5

Steel Trade Policy in Crisis

As the world steel market changed rapidly in the postwar period, the challenge to adjust fell heavily on established steel industries, first in the United States, then in the EC. It is largely the resistance to the market-driven adjustment process in the United States and the EC that has created the perpetual cycle of steel protectionism since 1968. The course of steel trade relations has seen three international 'policy crises' since that time, all of which orginated in protectionist sentiment in the United States. The first appeared in 1968, when American State Department officials concluded agreements whereby Japanese and EC steel producers would 'voluntarily' limit their exports to the United States. The second arose in 1977 with a rash of antidumping suits against Japanese and Community steel suppliers to the United States and culminated in the announcement of the introduction of the trigger price mechanism. The third grew out of the failure of earlier policies to provide relief to the American steel industry and effectively limit total imports to the United States, and led to a series of bilateral export restraint agreements with the EC in 1982 and with most other suppliers in 1984.

An examination of the development of import restrictions on steel during these 'crisis' periods reveals three unifying themes. The first is that protectionist sentiment arose as a result of periodic surges in exports from a small number of easily identifiable (and thus politically targetable) 'disrupters'. In the large American import market, Japan and the EC played this role, at times together, at times individually. The second theme emerges from the principal concern of policymakers during such 'crises'; how to mollify protectionist sentiment from an industry with such high political visibility and perceived strategic importance and simultaneously avoid spiralling retaliation. The third theme, finally, is found in the communicability and recurrence of the resulting government intervention in steel trade. Once a policy is implemented that isolates the 'disrupter', the trade disturbance tends to be refocused towards other markets, generating a similar policy response. The pattern in steel protectionism has been for the United

States, the largest and most open market in world steel trade, to experience an import surge and protectionist pressure first, followed by a repercussive protectionist policy on the part of the EC. At the same time, the failure of one set of protectionist policies to rescue the industry generates the bureaucratic momentum for new protectionist policies to bail out the first set.

The following three chapters set out to trace these developments and reveal the self-perpetuating nature of the discriminatory protectionist response of governments to the crisis. There is, in other words, a clear sequence and progression in the series of trade restrictions that have been used to protect the steel industries of the United States and the EC, from the first 'voluntary' export restraint agreements imposed by the United States to the first such agreements in the EC, followed by the trigger price mechanism in the United States and the implementation of a similar policy by the Community, leading, in turn, to the steel 'pacts' between the United States and most steel exporting countries. The analysis of this chapter begins with an overview of government involvement in the steel industry during the early postwar period; in many cases the role of government established at that time would go on to shape the actions of governments and steel firms during the more difficult years ahead. The first crisis situation is then examined in the background and initial consequences of the first VER agreements between the United States and its principal steel suppliers in Japan and the EC. These arrangements were important in shaping the policy-learning process for later instruments of 'voluntary' restraint, especially with regard to legal questions and specific product coverage.

Steel and the state: the early postwar years

In the wake of the Second World War, the steel industry's perceived importance as a strategic industry had become stronger than ever. The war had been won in large part on the overwhelming industrial strength of the United States, which had produced ships, aircraft, tanks, jeeps, arms and ammunition in unprecedented quantities. The traditional identification of basic industrial capacity and output as a measure of potential military capability was now etched clearly in the minds of government and would remain so for decades after the war. Steel, as the most basic of the smokestack industries, was now more than ever seen as the linchpin of a stable and secure economy, a role certainly too important to leave to steel firms or the private marketplace alone. All governments of the established steel-producing countries, as well as governments of countries aspiring to a position of industrial strength, therefore became involved in one way or another in the development of

their national steel industries in the postwar period. Given the diverging economic, historical, political and social environments in each country, however, the impact of government intervention was different in each case.

The United States: steel vs the government

The relationship between American steel companies and the US government in the early postwar period can be generally described as one of antagonism, a situation that distinguished the American steel industry from most other national industries at the time. For many years, Washington had cast a suspicious eye on the steel oligopoly, especially the US Steel Corporation, which had played the role of price leader since its foundation by merger in 1901. From the time of the annual 'Gary dinners' of 1907–11 (named for its president, Judge Gary, who used the occasions to encourage 'co-operative' pricing among the major steel firms) until the late 1950s, US Steel and the industry in general had been the subject of anti-trust investigations for their pricing practices. Unlike European steel industries, which also had oligopolistic or even cartelized market structures before the Second World War but had been practically destroyed during the conflict, the American steel industry entered the postwar period unscathed, retaining its market power and the continued scrutiny of the government.

In this atmosphere of distrust, US government officials began to focus for the first time on the peacetime steel capacity needs of the country. Their concern was a reflection of the general increase in government involvement in the economy since the depression years of the 1930s (carried through by presidential administrations of the Democratic Party until 1953) and of the apparent success achieved by the government in managing industrial production and prices during the war. Some government economists had concluded that American steelmaking capacity in the late 1940s was inadequate to serve an expanding American economy, and delivered official admonishments to steel executives to invest in new capacity. The fear among government officials was that, in the absence of adequate steel supply for the steel-consuming sectors, a bottleneck would develop, placing a drag on the entire economy (Barnett and Schorsch, 1983, p. 234). At the same time, those inclined to distrust the steel oligopoly anyway saw this as an opportunity for the industry to command premium prices and reap exorbitant profits. In his State of the Union message of 1949, President Truman suggested that the government might even take matters into its own hands, as he planned

To authorise an immediate study of the adequacy of production facilities of materials in critically short supply, such as steel; and if

found necessary, to authorise government loans for the expansion of production facilities to relieve such shortages, and furthermore to authorise the construction of such facilities directly if action by private industry fails to meet our needs.
(*New York Times*, 6 January 1949, p. 4)

The government's concern with adequate steel supply peaked during the Korean War, exacerbated by the continual threat of strikes. Labor-management relations in the early postwar period were anything but peaceful; strikes occurred in 1946, 1949, 1952, 1956 and 1959 (Lawrence, 1983, p. 71). The 1952 strike occurred during the Korean War; President Truman, once more digusted with what he considered a recalcitrant steel industry, responded by declaring a government takeover of the steel mills in order to maintain the war effort. Although the Supreme Court soon thereafter declared the seizure unconstitutional, it was becoming clear to the steel industry that its business had become extensively, and perhaps permanently, politicized.

The American steel industry did in fact expand capacity in the early 1950s, although the steel shortages of the Korean War years may have provided as much of an incentive as direct threats from government officials. Unfortunately, the expansion came at an inopportune time in terms of technological developments. All of the new capacity at this time consisted of open hearth furnaces, which, as noted in Chapter 4, became obsolete soon after it was installed as a result of the development of the basic oxygen furnace. These investments thus played a major part in the American steel industry's competitive decline. In so far as government policy played a role in the investment decision of steel firms during this period, it can thus rightly be judged a failure.

The major focus of government intervention in the steel industry during the early postwar period, however, was pricing. Increasing government involvement in the economy, coupled with the fear of postwar inflation, had created a particular concern with steel prices, which were seen as a critical determinant of the general price level (Barnett and Schorsch, 1983, p. 235). At the same time, the recognition of the steelworkers' union by US Steel in 1937 had introduced the complicating factor of labor negotiations every three years beginning in 1946 (P. Lawrence, 1983, pp. 70–1). The steel industry was loth to grant wage increases unless it could also raise steel prices; for a long-established and firmly entrenched oligopoly, the issue was one of 'fair' prices that would be established in this context through political leverage. This wage–price connection was the main concern of government officials in the 1950s and early 1960s, who participated in every labor-management negotiation during this period. Steel firms

were nonetheless successful in winning the government's *imprimatur* on price increases in exchange for wage increases granted to workers, which contributed to steel price increases in excess of the overall price level throughout the 1950s (Barnett and Schorsch, 1982, p. 235). The seemingly 'fair' exchange of wage concessions for price increases was, however, only possible in the absence of severe competitive pressures, and this practice would only continue as long as imports and domestic competition did not threaten the large intregrated firms' market share.

In the meantime, the steel industry's political lessons proceeded on the basis of hard experience, which seemed to show the futility of direct confrontation during the tri-lateral negotiations. A generous wage settlement in 1956, mediated by the government, had contributed to declining profits, creating industry resentment against both labor and government. When the industry balked at further concessions in 1959, the ensuing 116-day strike caused many steel customers to turn to foreign steel suppliers, making the United States a net steel importer. Declining conditions in the industry hardened steel executives' attitudes by the time of the 1962 wage negotiations. After the Kennedy Administration had secured a commitment to moderate wage increases by the steelworkers' union, US Steel defied the rules of the negotiating game by announcing an across-the-board price increase, an action that would normally have led to similar increases by other steel firms. However, President Kennedy immediately retaliated by (among other things) threatening to switch government contracts to steel firms that refused to follow US Steel's lead. A soft domestic market apparently aided the government's strategy; in any case US Steel rolled back prices three days later (McConnell, 1963).

The bitterness that characterized relations with the steelworkers' union and government after the war led the steel industry to two important conclusions that would play a significant role in trade policy developments. The first was that the industry's interests could best be asserted with strong political leverage in Washington; direct confrontations with the President were inadvisable. The industry's problem in this regard centered on the means by which broad-based political support could be achieved. The second conclusion was that there existed a potential commonality of interests between the industry and the union. Wage and price increases went together, and as long as the former could be kept in check, the union could act as a useful ally to achieve industry goals.

As imports began to impinge on the domestic steel market in the 1960s, the political battleground changed, and the industry's new political strategy incorporated the lessons of the early postwar period. Steelmakers could no longer administer prices designed to produce a 'fair' return, validated by government mediation of wage contracts.

This political constraint had been replaced with the economic constraint of increasingly competitive foreign steel supplies. The industry's attention therefore turned to the 'import issue' and to the legislative and administrative means by which import competition could be controlled. It was in this context that the support of the steelworkers' union was necessary, since congressional support, based on steelworker constituencies in key northeastern and midwestern states, would be essential to achieve or threaten direct import restrictions. The alliance of steel labor and management on this issue was sealed by the fact that imports threatened steelworker wages as well as steel industry profits.

Thus, the stage was set for the protracted battle over steel imports in the United States, in which the steel industry would exhibit considerable skill in attaining the requisite level of political influence to implement trade restrictions. Yet from an economic perspective the experiences of the postwar period that served as the basis of the industry's protectionist strategy only provided false lessons for the industry's long-term viability. By the time American steel firms began their competitive decline by ignoring the importance of cost-efficiency on *world* markets, they had internalized the principle of 'fair' prices that would somehow be validated and guaranteed by the government, regardless of the change in market conditions. The demands for import protection that would appear later were in this sense a continuation of the price negotiations of the 1950s. As the new competitive conditions inexorably closed in on the American steel industry, government intervention could only delay and draw out the painful process of adjustment.

Countries of the European Community

Like the United States, European countries emerged from World War II with a deep concern with national steel capacity and prices. Yet in contrast to the antagonistic relationship between government and industry in the United States, European government policies set out to support their national industries, although the effects of intervention were not uniformly beneficial. The reasons for the more supportive approach lay in large part in the condition of the industry and of the European economy in general at the end of the war. A large portion of steelmaking capacity lay in ruins or in severe need of modernization and the widespread destruction of the cities, infrastructure and manufacturing facilities required the reconstruction of basic industries such as steel before the rebuilding of the economy as a whole could proceed. Government policies therefore concentrated on promoting the construction of new steel plants from the ground up. In so far as economic

conditions seemed to favor the rebuilding of steel industries in these established steel-producing countries anyway, government aid in most cases probably did little to distort the competitive incentive structure for steel firms in the early postwar period. Furthermore, the establishment of the European Coal and Steel Community (ECSC) in 1952, which abolished trade barriers between the six original member countries (Belgium, France, West Germany, Italy, Luxembourg and the Netherlands), promoted competition within the Community and thereby provided competitive pressures to adjust, especially in the form of adaptation to new technologies. The dangerously misleading lesson governments seemed to draw from the success of early postwar reconstruction, however, was that government policies could also prevent or retard the decline of the steel industry when excess capacity and changing competitive conditions emerged two decades later.

Despite the generally supportive nature of government policies towards their national steel industries, the degree of intervention varied widely among the European countries. The steel industries of Belgium, Luxembourg and the Netherlands, for example, received little government aid in the early postwar period, although steel prices in these countries were subject to government control (Lister, 1960, p. 197). In West Germany, tax incentives encouraged investment in the steel industry, but little direct aid was granted at this time (ITC, 1984a, p. 68). In contrast, the French government pursued a much more activist role, guiding investment and production with the goal of increasing and modernizing steel capacity, but also imposed price controls on steel, which worked against these objectives. The close supervision of French steel industry activities by the state was in fact informally institutionalized by the common educational background of steel executives and civil servants, who both generally attend *l'Ecole Polytechnique*, the national polytechnical institute (DeWitt, 1983, p. 235). In Italy, as in France, government planning in the steel sector was also designed in part to aid the economic development of specific regions (Bollino, 1983, p. 267). In the United Kingdom, the level of government involvement depended in part on the political party in power at the time (see Ovenden, 1978). Previously the least interventionist of all steel-producing countries, Britain's Labour government nationalized most of the steel industry in 1950. The new Conservative government of the following year reversed this action, returning the industry to the private sector, but government involvement became firmly established in 1953, when a national Iron and Steel Board took on the responsibility of reviewing and approving steel industry activities and setting prices in consultation with steel producers (ITC, 1984a, p. 235). Labour Party legislation would then re-nationalize the industry in 1967.

Among the governments most directly involved in the steel industry, a pattern of public control was emerging, in which the steel industry was to serve broad economic, social and regional goals. The idea of the steel industry as a servant to the public interest was not a new one, and had ironically appeared in the early public acclamation for the International Steel Cartel, despite the fact that this collusive arrangement had done great damage to consumer welfare. All of the major European steel-producing countries had participated in the ISC, and even though the principles of economic liberalism and anti-trust tended to dominate postwar economic policies, there was a lingering fascination among governments with the possibilities of 'co-operative' production and pricing agreements under public control.

These opposing principles of economic policy were combined in the Treaty of Paris of 1952, which established the ECSC. The treaty's general provisions seem forthright in their prohibition of any agreements in restraint of trade. Article 65 forbids 'all agreements among enterprises, all decisions of associations of enterprises, and all concerted practices, tending directly or indirectly, to prevent, restrict or distort the normal operation of competition within the common market' and goes on to enumerate the forbidden practices of price-fixing, production controls and market-sharing arrangements (Lister, 1960, p. 177). At the same time, this provision does not prohibit individual governments from interfering with the market, although such government actions would presumably not be allowed to interfere with trade within the ECSC. This interpretation thus validated the various forms of government planning and control over national steel industries at the time. The most significant deviation from market principles appears in the 'crisis' provisions of articles 58, 61 and 74, however. When such 'crisis' conditions appear in the ECSC, the supra-national authority may intervene directly in the market by setting voluntary or mandatory minimum prices (article 61) as a first step, or in the case of 'manifest crisis' by imposing production quotas (article 58) and restrictions on imports from third countries (article 74c). The definition of 'crisis' in these articles has been a matter of debate (see Joliet, 1981). In any case, the treaty leaves the door open to the official imposition of the same cartel arrangements that it set out to abolish.

Thus, the spirit of the defunct International Steel Cartel lived on in the crisis provisions of the ECSC. The reasoning behind their presence in the Treaty of Paris appeared to be that the problem with industrial cartels was merely that decision-making was in private hands, and was therefore not answerable to the public interest. Public control of cartel arrangements, it was thought, could best solve the problems of fluctuating prices and overcapacity, while simultaneously protecting consumer and other national interests. This bureaucratic machinery

would later be activated when EC officials established a new cartel in 1977.

The seeds of conflict over steel trade

In general, developments in the relationship between the steel industry and government had placed trade policy in the United States and the EC on a collision course with the stated doctrine of open trade adopted in the postwar period. In the United States, an entrenched steel oligopoly had adopted the strategy of pursuing political channels to achieve its goals, and was therefore in a strong political position to resist adjustment pressures by securing trade restrictions. In the EC, the growing trend towards governments' use of the steel industry as a means to pursue national goals, reinforced by 'crisis' contingency plans of the ECSC, indicated the likelihood of massive government control over the steel market, including trade restrictions, if adjustment pressures threatened the industry. Against this political background, the changing competitive structure of the world steel market was to trigger a protracted spiral of steel protectionism.

Origins of the first crisis

The United States, with its large steel market, became a net importer of steel in 1959. The growing level of import penetration in the following decade made the American domestic market increasingly sensitive to world market developments, thereby causing it to become the primary battleground of protectionism in steel. By regulating access to the world's largest steel import market, American trade policy came to affect profoundly trade conditions in and among steel exporters worldwide.

On the other hand, the greatest supply-side influence in world steel trade during this period can be traced to Japan, which burst upon the world market as a major steel exporter. As a result, world export market shares shifted from West European countries and the United Kingdom to the newcomer, Japan — a trend which was also reflected in the composition of United States imports (see Table 5.1).

It was thus increased import market penetration, led by the Japanese, which set the stage for the protectionist debate in the United States in 1967 and 1968. The call for import restrictions was prompted by a sharp drop in domestic demand, caused in part by cyclical market forces, in part by the threat of a steelworkers' strike which drove major steel consumers to foreign suppliers. Soft domestic demand heightened the visibility of increased import penetration, and a concomitant fear of increased dependence on foreign steel, of increased unemploy-

Table 5.1 *Imports into the United States of steel mill products by region of origin ('000 net tons)*

	Japan (%)		European Community (%)		Other (%)		Total (%)	
1961	597	(18.9)	1,952	(61.7)	614	(19.4)	3,163	(100)
1962	1,072	(26.1)	2,087	(50.9)	941	(23.0)	4,100	(100)
1963	1.808	(33.2)	2,246	(41.2)	1,398	(25.6)	5,452	(100)
1964	2,446	(38.0)	2,585	(40.1)	1,408	(21.9)	6,439	(100)
1965	4,418	(42.6)	4,191	(40.4)	1,774	(17.0)	10,383	(100)
1966	4,851	(45.1)	3,841	(35.7)	2,061	(19.2)	10,753	(100)
1967	4,468	(39.0)	4,842	(42.3)	2,145	(18.7)	11,455	(100)
1968	7,294	(40.6)	7,097	(39.5)	3,569	(19.9)	17,960	(100)
1969	6,253	(44.6)	5,199	(37.0)	2,582	(18.4)	14,034	(100)
1970	5,935	(44.4)	4,573	(34.2)	2,856	(21.4)	13,364	(100)
1971	6,908	(37.7)	7,174	(39.2)	4,240	(23.1)	18,322	(100)
1972	6,440	(36.4)	7,779	(44.0)[a]	3,462	(19.6)	17,681	(100)
1973	5,637	(37.2)	6,510	(43.0)	3,003	(19.8)	15,150	(100)
1974	6,159	(38.6)	6,424	(40.2)	3,387	(21.2)	15,970	(100)
1975	5,844	(48.6)	4,118	(34.3)	2,050	(17.1)	12,012	(100)
1976	7,984	(55.9)	3,188	(22.3)	3,113	(21.8)	14,285	(100)
1977	7,820	(40.5)	6,833	(35.4)	4,654	(24.1)	19,307	(100)
1978	6,487	(30.7)	7,463	(35.3)	7,185	(34.0)	21,135	(100)
1979	6,336	(36.2)	5,405	(30.9)	5,777	(32.9)	17,518	(100)
1980	6,007	(38.8)	3,887	(25.1)	5,601	(38.1)	15,495	(100)
1981	6,220	(31.2)	6,482	(32.6)	7,196	(36.2)	19,898	(100)
1982	5,185	(31.1)	5,597	(33.6)	5,881	(35.3)	16,663	(100)
1983	4,237	(24.8)	4,114	(24.1)	8,719	(51.1)	17,070	(100)

Source: Annual Statistical Report, American Iron and Steel Institute, Washington, various issues.

[a] From 1972, figures for the European Community include those of the United Kingdom, which joined the Community's VRA with the United States in that year.

ment among the well-organized steelworkers and of the impact of steel imports on the balance of payments heightened protectionist resonance among legislators (US Congress, 1968, pp. xxv–xxix). Import-competing firms in the United States took advantage of the protectionist mood by pursuing legislative and trade-statutory avenues of protection simultaneously, a pattern which was to be repeated under more severe circumstances in 1977 and 1984. The first approach took the form of lobbying by the steel industry in Congress for quantitative restrictions on steel imports with the aid of the steelworkers' union. These efforts culminated in the introduction of a bill proposing com-

prehensive 5-year bilateral import quotas on all supplying countries.[1] The bill provided a credible protectionist threat since it had thirty-five co-sponsors in the Senate and enjoyed broad-based congressional support (Lowenfeld, 1979, p. 197). The second channel involved a complaint filed with the Treasury Department by the United States Steel Corporation against France, the Federal Republic of Germany, Italy, the Netherlands, Belgium and Luxembourg, alleging subsidization of steel exports and calling for the imposition of countervailing duties (*New York Times*, 10 October 1968, p. 1). The complaint focused on the EC's practice of remitting the value added tax (VAT) on export sales.

The executive branch of the United States government (then under President Lyndon Johnson) opposed the use of these instruments of trade restriction for several reasons. Aside from the foreign retaliation which would result were they implemented, trade officials knew that they would jeopardize the credibility of American commitments to international trade agreements. The Administration's special trade representative had recently participated in the negotiations which achieved the harmonization of steel tariffs during the Kennedy Round of multilateral trade negotiations and the progress made in those talks would have been completely eliminated by the proposed non-tariff barriers.[2] Furthermore, the quota proposal was in direct violation of Article XI of the General Agreement on Tariffs and Trade (GATT), which proscribed quantitative import restrictions. As for the subsidy complaints, they were in apparent violation of GATT Article VI(4) and would elicit swift and severe foreign retaliation if sustained (Lowenfeld, 1979, pp. 188–90). In general, United States trade officials were concerned that the pursuit of traditional protectionist remedies to aid American steel producers would unravel the delicate trade liberalization measures which had been so painstakingly negotiated, most recently during the Kennedy Round negotiations.

At the same time, the apparent penetration of Congress by protectionist sentiment was perceived by the executive branch to indicate that some government effort to satisfy the demands from industry for import restrictions was necessary if passage of the quota bill was to be avoided. Within the executive branch anticipation of the threatened protectionist legislation convinced officials there that an intermediate protectionist device was needed to forestall the implementation of the original, more severe proposal. Yet what alternative policies were available? International trade commitments, as mentioned, eliminated consideration of tariffs and unilaterally imposed quotas even in watered-down form. The questionable use of American trade laws to impose countervailing duties against VAT rebates was equally unpalatable. Temporary escape-clause relief under Article XIX of the

GATT would also be difficult to justify, since the American steel industry's problems could not be attributed to trade agreement concessions (Lowenfeld, 1979, p. 194).

In the end, the circumstances surrounding the choice of an alternative policy were shaped largely by anxieties among steel exporters to the United States equal to those of American trade officials. Japan, in particular, was worried that its status as a 'disrupter' would lead to severe bilateral trade restrictions under the proposed quota bill. The concern of Japanese steel exporters led Yoshihiro Inayama, Chairman of the Japan Iron and Steel Exporting Association, to propose 'voluntary' export restraint (VER) as a means of neutralizing protectionist sentiment in the United States.[3] The steel exporters in the EC agreed to this approach, and discussions over the size and distribution of the restraint began in March 1968. In the meantime, the quota bill was shelved (Lehnardt, 1969, p. 109).

The VER negotiations in 1968 revealed a convergence of interests among foreign producers, import-competing American producers and American trade policy officials.[4] For the foreign exporters convinced of the strength of protectionist sentiment in the United States, 'voluntary' restraint was perceived as the best arrangement they could achieve under the circumstances. Not only would a VER policy allow them to participate in setting their own export limit (a role they would have forfeited under a unilaterally imposed quota), it would also allow them to impute the scarcity premium of the restricted exports to themselves. For American producers, a policy of VER represented an alternative perhaps inferior to the more severe import curb expected from direct import quotas, but it still offered a desired quantitative restriction at low political cost and risk. For American trade policy officials in the executive branch, the VER approach represented a compromise which could satisfy protectionist sentiment and defuse protectionist legislation simultaneously, accomplishing a trade restriction without alienating trading partners and without violating the letter of the GATT (see Jones, 1984).

Nevertheless, while the prospect of 'voluntary' export restraint provided a convenient and politically expedient alternative to traditional trade policy instruments as a solution to the steel import 'crisis', it was clear from the outset that the negotiations would venture into uncharted legal waters. The primary concern in this regard was the status of such an agreement under United States anti-trust laws. Normally, any conspiratoral agreement among firms which fixes prices or restricts supply violates Article 1 of US Code 15, the Sherman Anti-trust Act. The obsession among participants in the agreement with establishing immunity from anti-trust violations was reflected in the structure of the negotiations. The State Department acted as in-

termediary during the talks, which included Japanese and Community steel producers, representatives of American steel firms (who could thereby bargain indirectly with the steel exporters) and members of the United States Congress. In this manner, a direct collusive (and thus illegal) agreement among firms could, supposedly, be avoided. In addition, the State Department advised Japanese and European steel makers to draft their acceptance of the agreement in the form of 'statements of intent', emphasizing the supplicatory nature of the negotiations and the non-binding character of the agreements. It was hoped that this legal format would provide further insurance against anti-trust violations. None the less, producers in the EC made sure to submit preliminary drafts of the final agreement to the United States Justice Department of an assurance of immunity which was, apparently, given (Lehnardt, 1969, p. 110).

While such legal questions seemed to command attention during the negotiations haggling over substantive issues lasted for several months (Lehnardt, 1969, p. 109). Agreement on the tonnage amounts and distribution and on the time framework of the restraints was finally reached, however, and the Japanese and European 'statements of intent', dated 23 December 1968, arrived at the State Department. These were turned over to the congressional Ways and Means Committee and the Senate Finance Committee for an official announcement on 14 January 1969 of the 'voluntary restraint agreements'. The decision to have the formal announcement made through congressional channels appears to have been intended to lend the *imprimatur* of legitimate trade policy giving authority to the arrangements (Lowenfeld, 1979, p. 203).

Content and performance of the first agreements, 1969–71
The novelty of the VRAs and the attendant pre-occupation with their status under United States anti-trust laws evidently made negotiators hesitate to risk making the agreements too detailed. The VERs consisted, in fact, of a simple promise not to exceed given export levels of total steel mill products (by weight) to the United States from 1969 to 1971.[5] Specifically, the VRAs contained the following provisions:

(1) An 'assurance' on the part of Japanese and European Community steel exporters to limit shipments of steel mill products to the United States to 5.75 million net tons each during the 1969 calendar year. It is important to note that, although the negotiations included only the nine largest Japanese producers and six national producers' associations in the ECSC, the tonnage limits applied to *all* producers in Japan and the European Community,

including those smaller exporters not under direct control (see Lowenfeld, 1979, pp. 205–8).

(2) A 5 per cent annual growth allowance in shipment tonnage limits to the United States in 1970 and 1971;

(3) A promise to maintain the product mix and pattern of distribution of trade during 1969–71; and

(4) A statement basing the agreements upon three assumptions: (a) that total world exports to the United States would not exceed 14 million tons in 1969 (with 5 per cent growth allowances for 1970 and 1971); (b) that the United States would not, unilaterally, increase import restrictions on steel from the respective VRA partners during this period; and (c) that the VRAs would not violate American laws and that they conformed to international laws.

Table 5.2 shows how the VRA 'assurances' operated in practice. Import-competing steel firms in the United States were primarily interested in reducing import penetration levels and, in 1969, the VRAs appeared to be successful in this regard. Total imports of steel mill products in that year dropped to the target level of 14 million tons, down from the 1968 figure of 18 million tons, with a corresponding drop in import penetration ratio of 16.7 per cent to 13.7 per cent. In 1970, imports dropped still further to 13.4 million tons, with import penetration at 13.8 per cent. An examination of steel trade patterns also shows that exports as a whole were apparently diverted from the United States to other markets in the first year of the VRAs, with the greatest shifts re-directing Japanese exports towards the EC and Community exports towards the countries of the European Free Trade

Table 5.2 *Deliveries of Steel Mill Products and VRA Compliance (I), 1969-71*

Year	Actual imports ('000 net tons)				Imports VRA Ceiling (%)	
	Japan	European Community	Other	Total	Japan	European Community
1969	6,253	5,199	2,582	14,034	109	90
1970	5,953	4,573	2,856	13,364	98	72
1971	6,908	7,174	4,242	18,324	109	113

Note: Adapted from a study for the FTC (1977), and from statistics of the US Department of Commerce, Bureau of Resources and Trade Assistance and Office of Import Programs.

Association (EFTA) (see Tables 5.3 and 5.4). In 1971, however — the final year of the agreements — steel imports to the United States rose sharply to 18.3 million tons, representing an import ratio of 17.9 per cent, the highest American steel import figures recorded to that date. Imports from the Community and Japan exceeded their VRA limit by 1.5 million tons.

The checkered performance of the first VRAs can be traced in part to the diminishing credibility of their threatened alternative (quotas) and in part to changing market conditions unrelated to the implementation of the agreements. The first year of the agreements marked the height of foreign suppliers' concern that the United States would enact unilateral protectionist legislation. Compliance with the VRAs was thus seen as preferable to the risk presented by the threat of severe

Table 5.3 *Japanese steel exports by country of destination, 1963–82 ('000 metric tonnes)*

Year	Total	United States (%)		European Community (%)[a]		Far East (%)	
1963	5,283	1,544	(29.2)	464	(8.8)	1,825	(34.5)
1964	6,539	2,352	(35.8)	298	(4.6)	2,139	(32.7)
1965	9,746	4,122	(42.3)	160	(1.6)	2,502	(25.7)
1966	9,478	4,416	(46.6)	268	(2.8)	2,867	(30.2)
1967	8,707	4,094	(47.0)	181	(2.1)	2,851	(32.7)
1968	12,774	6,617	(51.8)	210	(1.6)	3,609	(28.3)
1969	15,548	5,272	(33.9)	947	(6.1)	4,710	(30.3)
1970	17,589	5,580	(31.7)	950	(5.4)	5,314	(30.2)
1971	23,194	5,787	(25.0)	1,653	(7.1)	6,819	(29.4)
1972	20,922	5,658	(27.0)	1,116	(5.3)	6,666	(31.9)
1973	24,805	4,696	(18.9)	1,282	(5.2)	9,629	(38.8)
1974	32,220	5,791	(18.0)	1,054	(3.3)	10,869	(33.7)
1975	28,942	5,126	(17.7)	1,573	(5.4)	8,778	(30.3)
1976	36,016	6,800	(18.9)	1,467	(4.1)	10,722	(29.8)
1977	33,628	7,016	(20.9)	1,265	(3.8)	12,235	(36.4)
1978	30,925	5,614	(18.2)	623	(2.0)	14,068	(45.5)
1979	30,697	5,799	(18.9)	720	(2.3)	13,306	(43.3)
1980	29,705	4,883	(16.4)	616	(2.1)	12,959	(43.6)
1981	28,455	5,920	(20.8)	198	(0.7)	11,458	(40.3)
1982	28,635	3,913	(13.7)	260	(0.9)	12,020	(42.0)

Source: United Nations Economic Commission for Europe, *Statistics of World Trade in Steel,* various issues.

[a] From 1973, the figures for the European Community include those of the United Kingdom, Denmark and the Republic of Ireland.

Table 5.4 *European Community steel exports by country of destination 1963–83 ('000 metric tonnes)*

Year[a]	Total	United States (%)	EFTA (%)	Asia (%)	Eastern Europe (%)	Africa (%)
1963	9,063	1,467 (16.2)	3,092 (34.1)	1,115 (12.3)	575 (6.3)	780 (8.6)
1964	10,490	1,826 (17.4)	3,610 (34.4)	1,116 (10.6)	415 (4.0)	887 (8.5)
1965	14,290	3,404 (23.8)	3,540 (24.8)	1,602 (11.2)	379 (2.7)	1,238 (8.7)
1966	12,400	2,891 (23.3)	3,257 (26.3)	1,598 (12.9)	402 (3.2)	910 (7.3)
1967	14,322	3,814 (26.6)	3,340 (23.3)	1,970 (13.8)	926 (6.5)	958 (6.7)
1968	15,345	5,620 (36.6)	3,616 (23.6)	1,632 (10.6)	898 (5.9)	928 (6.0)
1969	14,143	3,822 (27.0)	4,495 (31.8)	1,314 (9.3)	962 (6.8)	928 (6.6)
1970	13,463	3,599 (26.7)	4,185 (31.1)	1,274 (9.5)	800 (5.9)	1,149 (8.5)
1971	16,210	5,852 (36.1)	4,144 (25.6)	1,356 (8.4)	894 (5.5)	1,257 (7.8)
1972	17,544	5,234 (29.8)	4,917 (28.0)	1,565 (8.9)	1,282 (7.3)	1,351 (7.7)
1973	18,294	4,349 (23.8)	3,538 (19.3)	2,487 (13.6)	2,647 (14.5)	1,882 (10.3)
1974	24,289	4,910 (20.2)	4,055 (16.7)	2,728 (11.2)	4,202 (17.3)	2,841 (11.7)
1975	20,815	2,887 (13.9)	3,480 (16.7)	2,775 (13.3)	3,716 (17.9)	2,702 (13.0)
1976	16,474	2,711 (16.5)	3,261 (19.8)	2,155 (13.1)	3,372 (20.5)	2,226 (13.5)
1977	21,497	6,213 (28.9)	3,318 (15.4)	2,748 (12.8)	2,688 (12.5)	2,538 (11.8)
1978	25,770	5,726 (22.2)	3,413 (13.2)	6,898 (26.8)	3,376 (13.1)	2,541 (9.9)
1979	24,708	4,623 (18.7)	3,721 (15.1)	5,970 (24.2)	3,520 (14.2)	2,529 (10.2)
1979	24,708	4,623 (18.7)	3,721 (15.1)	5,970 (24.2)	3,520 (14.2)	2,529 (10.2)
1980	22,189	2,904 (13.1)	4,234 (19.1)	4,251 (19.2)	3,215 (14.5)	3,033 (13.7)
1981	23,499	4,338 (18.5)	4,151 (17.7)	4,583 (19.5)	2,550 (10.9)	2,851 (12.1)
1982	18,349	3,448 (18.8)	3,741 (20.4)	3,443 (18.8)	2,152 (11.7)	2,043 (11.1)
1983	18,833	3,438 (18.3)	3,940 (20.9)	4,475 (23.8)	2,208 (11.7)	1,952 (10.4)

Source: Iron and Steel Yearbook, Statistical Office of the European Community, Luxembourg, various issues.

[a] The figures for 1963–73 are for the six original member countries of the European Community; from 1974, figures are for the Community of the nine countries.

import restrictions. It is also probable, however, that demand for imports among American steel consumers declined independently in 1969 as a result of the steelworkers' contract settlement, which had prompted hedge buying abroad in 1968. In 1969, a significant portion of the drop in imports in the United States may thus have been caused by changing risk conditions associated with supply sources, rather than by the VRA quotas.

In 1970, domestic prices in the United States, the EC and Japan all rose and import penetration of the United States market did not change appreciably, as compared with 1969. 1971, however, saw a general downturn in domestic price levels in the Community and Japan, while American prices continued to rise (FTC, 1977, p. 221).

Market opportunities thus encouraged increased import penetration in that year, leading to deliveries from both the Community and Japan in excess of their VRA limits. By this time, the low import penetration levels and the 'good' performance of the VRAs had reduced protectionist sentiment in the United States and, thereby, its credibility as a means of enforcement. In the absence of an effective mechanism to induce compliance, 'voluntary' restraint was overridden by market forces. Furthermore, 'voluntarily' restrained suppliers, anticipating new negotiations, sought to establish the highest possible import levels in the final year of the agreement.

Another possible channel of market force erosion of the VRAs can be found in the fact that not all exporting firms in Japan and the EC were covered by the agreements. Unrestrained export deliveries from smaller firms not party to the VRA evidently caused especially serious concern among the nine large Japanese firms participating in the agreement, since they controlled only 85 per cent of export suppliers.[6] Even in the first apparently 'effective' VRA year, Japanese exports to the United States exceeded their limit and were offset by 'honest', but angry, Community producers. While this 'disruption' of the agreed VRA parity between deliveries from the Community and Japan could have been caused by the secular competitive trend of increased Japanese production efficiency and import penetration, the control of non-participating exporting firms in Japan remained a serious threat to the effectiveness of the VRAs.

A final problem of the VRA in achieving its intended purpose emerged from its simple designation of export limits by weight, with no product mix specification. This allowed exporters to shift their emphasis to the sale of more expensive alloy and specialty steels, allowing them to keep within the weight limit while maintaining the value of their exports. Thus, from 1968–70, total steel imports in the United States decreased *in weight* by 25 per cent but the *value* of steel imports declined only 0.5 per cent, from $1.976 billion in 1968 to $1.967 billion in 1970. Expensive alloy tool steel imports, for example, increased 29 per cent during the period, from 13,453 net tons to 17,356 net tons (FTC, 1977, p. 74). Any such shift in trade attributable to the VRAs represented a sort of trade diversion by product mix, causing resources to be misallocated among the various steel product manufacturing processes. The more serious political problem it presented to American trade policymakers, however, was possible import disruption refocused on domestic specialty steel manufacturers.

In summary, an overall view of the trend in import penetration suggests that VRAs may have had some restraining effect on American steel imports as long as a credible protectionist threat was maintained. Market forces seemed, however, to be moving towards a

similar reduction in import penetration in 1969 and 1970, and any restraining influence of the VRAs apparently disintegrated by 1971, when imports surged above their limits. The major shortcomings of the VRA as a means of effective import restriction included: (1) the lack of an enforcement system; (2) the failure to include all potential exporters in the agreement; and (3) the absence of specific coverage by detailed product groupings.

Renegotiation and renewal

In spite of the dubious performance of the initial VRAs, major American steelmakers seemed satisfied with their apparent success in 1968 and 1969 and sought a renewal of the agreements, with the understanding that new measures for enforcement and more specific product coverage would be added. State Department officials were emboldened by the apparent legal viability of the first agreements as an instrument of trade policy and began negotiations for more detailed VRAs in May 1971. Steel exporters in Japan and the EC were willing to conclude new agreements as a means of keeping United States protectionist legislation in check. In addition, the VRA continued to offer, in principle, a generous accommodation of producer interests: access to the large American import market free of further trade barriers, with the ability to gain scarcity rents if the restraints were effective.

The second round of VRA negotiations, covering exports to the United States from 1972–4, were concluded in May 1972.[7] The basic format of the agreements remained the same, but their provisions were much more detailed. In addition, the participation of the United Kingdom, a major steel exporter, in the 1972 VRA, as a new member of the EC, had to be factored into the re-negotiated export limits. The basic components of the new VRA were:

(1) voluntary limits on steel exports to the United States in 1972 of 7.27 million metric tonnes from the European Community and the United Kingdom and of 5.895 million metric tons from Japan;

(2) annual growth rate allowances on these limits, of 1 per cent in 1973 and 2.5 per cent in 1974, for the Community and the United Kingdom, and of 2.5 per cent in 1973 and 1974, for Japan;

(3) an assurance that no more than 60 per cent of the total annual shipment (by weight) would be delivered in either semester (January/June or July/December) of that year;

(4) specific voluntary limits on stainless, alloy and high speed steel mill products for each of the years 1972–4;

(5) voluntary target, export growth-rate limits for cold finished steel bars and fabricated structural steels: 2.5 per cent over preceding-year shipments in 1973 and 1974;

(6) an assurance that, for the years 1972−4, the basic geographic mix of delivery destinations by customs region (Atlantic, Gulf Coast, Pacific and Great Lakes) and volume would not differ from the pattern set from 1969−71;

(7) a provision for consultations on problems arising from the agreement; and

(8) a statement basing the agreement on the assumptions that (a) the VRA participants would not lose market shares to non-participating countries as a result of the agreement, (b) the United States would not increase trade barriers on steel against VRA participants unilaterally and (c) the agreement was not in violation of United States law or international rules.

As compared with the first VRAs, the second agreements envisaged a much more stringent and comprehensive control of steel trade. Many of the changes and additions included in the new VRAs reflected a desire among American steel makers to correct the shortcomings revealed in the operation of the 1968 agreements. The specific limits on specialty steel exports, for example (item 4 above), were designed to stop the shift towards trade in higher-valued steels which occurred during the period of the first VRAs. In addition, the attention given to differentiated products (items 4 and 5), to the distribution of shipments during the year (item 3) and to the geographic concentration of imports (item 6) provided the opportunity to monitor more closely specific breaches of 'voluntary' restraint, even though no formal enforcement mechanism was mentioned.

The one motivating theme which emerges from an examination of the second VRAs, however, is the prevention of market disruption. Initial restraint levels were set below the record 1971 figures and annual export growth allowances (item 2) were set below those of the first VRAs (1 per cent to 2.5 per cent, as compared with 5 per cent). In the case of some specific products, the relatively low 1970 import levels were used as a base (item 5). The induced restraint in the time and the geographic destination and distribution of exports (items 3 and 6) was intended to prevent disruptive surges in deliveries and regional domestic market share erosion to imports, respectively. The provision for consultations (item 7) would allow for *ad hoc* settlements of unforeseen market 'problems'. These import restrictions clearly went well beyond the scope of traditional trade policy instruments, imposing a rigid structure of import market organization upon the modest foundation of the 1968 agreements. In short, the 1972 VRAs aspired, by all appearances, to a sort of informal orderly marketing agreement encompassing world steel exports to the United States.

Ironically, market developments, especially with regard to exchange rates, independently reduced United States imports from

1972–4 rendering superfluous all the tedious negotiations and agreement details (Table 5.5). The devaluation of the dollar in 1971 and 1973 caused all American import prices to rise and high domestic demand in Japan and the EC reduced steel export supplies (see Comptroller General, 1974, p. 18). In general, the dollar value of domestic steel prices in all major producing countries rose sharply during the period, reflecting a worldwide boom in the demand for steel. Firms in the United States maintained production during this time at an average capacity utilization ratio of 92.4 per cent (compared with 84.2 per cent from 1969–71) (Council on Wage and Price Stability, 1977, p. 145). These market conditions boosted the American industry to its most profitable performance in several years (see Figure 4.1) and import penetration levels decreased to 16.6 per cent in 1974. Thus, the apparent 'compliance' of VRA participants with their restraint quotas from 1972–4 was effected, to a large extent, by the dictates of the market. Both United States government officials and domestic industry representatives agreed with this assessment (Comptroller General, 1974, p. 18).

In summary, the renewed VRAs of 1972 succeeded in introducing measures to prevent market disruption and tighten product coverage not included in the first agreements. They appeared, on paper, to offer a much more comprehensive set of trade restrictions. In addition, the 1972 VRAs included the United Kingdom, the largest exporter to the United States not covered by the first VRAs. Their influence on trade flows, however, was at best marginal. As a result of market forces favorable to American steelmakers, the agreements were not really put to the test of crisis conditions.

The legal challenge

While market conditions had cast doubt upon the effectiveness of 'voluntary' restraint in regulating steel exports to the United States, a civil suit brought against State Department officials and firms participating in the 1972 VRA cast doubt upon its status under American constitutional and anti-trust laws. The suit was filed by the Consumers' Union, a non-profit public-interest agency, on 24 May 1972, shortly after the renewed VRAs were concluded.[8] The ensuing litigation lasted 3 years. Although the final court decision sustained the defendants, it raised serious questions as to the compatibility of export restraint agreements with anti-trust law.

While a detailed analysis of the case would go beyond the scope of this study, an examination of its salient points reveals its most important policy implications. The original suit contained two charges: (1) that the VRAs stood in violation of the Sherman Anti-trust Act and (2) that the conclusion of the VRAs represented an unlawful action by

Table 5.5   *Deliveries of Steel Mill Products and VRA Compliance (II), 1972–4*

	Actual imports ('000 net tons)				Imports/VRA ceiling (%)	
Year	Japan	European Community	Other	Total	Japan	European Community
1972	6,440	7,779	3,462	17,681	99	97
1973	5,637	6,510	3,003	15,150	85	80
1974	6,159	6,626	3,387	15,970	90	77

Note: Adapted from a study for the FTC (1977) and from statistics of the US Department of Commerce, Bureau of Resources and Trade Assistance and Office of Import Programs.

[a] Figures for the European Community include the United Kingdom, the largest exporter to the United States not covered by the first VRAs.

State Department officials to restrict trade without the procedural requirements of the Trade Expansion Act of 1962 (see *Consumer Reports*, August 1972, pp. 528–31). The Consumers' Union, however, subsequently decided that it did not have the funds that would be needed to pursue an anti-trust suit, so it withdrew part (1) of its complaint.

Thus, the most significant legal question regarding the VRAs – whether they represented a 'conspiracy in restraint of trade' – was cast aside, and the Consumers' Union based its case on the more technical charge that State Department officials acted in excess of their legal power. The plaintiffs' argument rested essentially on provisions of the Trade Expansion Act of 1962 which state that, before the United States government negotiates an agreement restricting foreign imports, the Tariff Commission must first conduct an investigation to determine whether imports of the product in question are causing or threatening serious injury to import-competing domestic producers. The defendants claimed that State Department officials acted lawfully under the aegis of the President, whose constitutional powers over foreign affairs allowed him to conclude such agreements. If the President acted lawfully, their argument continued, then any private parties involved could not be held liable (Lowenfeld, 1979, pp. 216–17).

The District Court finally sustained the defendants, ruling that the Trade Expansion Act did not restrict the President from negotiating with foreign firms on commercial matters. The Court also stated, however, that, although the amended complaint by the Consumers' Union excluded the anti-trust allegation, the VRAs were indeed *not*

immune from anti-trust prosecution. While this statement had no bearing on the Court's decision, it brought the ultimate legality of the agreements into question (Lowenfeld, 1979, p. 217).

Under this cloud of uncertainty, the Consumers' Union appealed against the decision. The United States Court of Appeals upheld the District Court's decision, however, ruling that the VRAs represented only a precatory, and not a legal, instrument of trade restriction and thus did not fall under the provision of the Trade Expansion Act.[9] Yet in spite of the Court decisions favorable to the VRAs, the Consumers' Union suit remained a great source of apprehension for those involved in the VRA negotiations, particularly government officials, until the Trade Act of 1974 defined the executive branch's authority to negotiate such agreements. The compatibility of such agreements with the principles of anti-trust remained in doubt, however, as shown by scholarly and legal criticism of the jurisprudence in the case (Lowenfeld, 1979, p. 230).

The boom market in steel in 1974 provided, therefore, a convenient excuse for State Department officials to allow the VRAs to expire quietly without further renewal. In the end, the Consumers' Union lawsuit demonstrated the tenuous legal status of an explicit, openly negotiated agreement restricting market supply. This lesson would help to shape policy thinking when methods of induced 'voluntary' restraint again came under consideration.

Policy children of the VRAs

It was noted earlier that, while the VRAs did not apparently alter the relative import shares of the EC and Japan in the American market, they did divert steel trade as a whole to other import markets. The main policy effect of this phenomenon was to refocus import pressure from the United States to the world's next-largest steel importing region, the Community. At the same time, some exports from the EC were diverted to other West European markets, particularly those in EFTA countries. The EFTA, however, has no common import policy and thus could not mount a co-ordinated protectionist policy response. From 1968–9, Japanese steel deliveries to the Community increased by 173 per cent and rose by another 67 per cent in 1970 (Table 5.6). Although the original import base was small, the trend in Japanese import penetration alarmed steel producers in the Community, who attempted to conclude VRA-type arrangements with Japanese exporters. There were, however, fundamental differences in the policymaking environment of the Community that prevented the negotiation for VERs by Community officials. During the period of the VRAs in the United States, the Community, as a supranational organization, had

Table 5.6   *Deliveries of Japanese steel to the European Community, 1961–83 ('000 metric tonnes)*

Year	Total	Japan	%	Year	Total	Japan	%
1961	1,909	2	0.0	1973	6,028	979	16.2
1962	2,461	141	5.7	1974	4,198	583	13.9
1963	3,316	478	14.4	1975	6,145	1,548	25.2
1964	2,676	280	10.5	1976	9,768	1,650	16.9
1965	1,905	134	7.0	1977	9,949	1,669	16.9
1966	2,268	242	10.7	1978	9,856	773	8.7
1967	2,626	158	6.0	1979	9,416	601	6.4
1968	2,947	221	7.5	1980	8,992	562	6.3
1969	4,970	604	12.2	1981	6,566	164	2.5
1970	6,749	1,011	15.0	1982	8,736	237	2.7
1971	5,163	993	19.2	1983	8,491	255	3.0
1972	6,586	1,297	19.7				

Source: Iron and Steel Yearbook, Statistical Office of the European Community, Luxembourg, various issues.

developed neither the authority nor the bureaucratic structure to co-ordinate and carry out sectoral trade negotiations or restrictions in the steel industry. In addition, import penetration in Community countries was still low enough to prevent general protectionist sentiment against Japanese steel from reaching the threshold level necessary for individual member countries to consider trade restrictions along the lines 'voluntary' restraint (see Table 5.7).

Table 5.7   *Import penetration in the European Community, 1967–83 ('000 metric tonnes)*

Year	Apparent total consumption	Imports	%	Year	Apparent total consumption	Imports	%
1967	57,618	2,626	4.6	1976	98,961	9,768	9.9
1968	65,105	2,947	4.5	1977	94,195	9,949	10.6
1969	76,507	4,970	6.5	1978	92,579	8,956	9.6
1970	77,337	5,749	8.7	1979	99,525	9,416	9.5
1971	72,034	5,163	7.2	1980	94,586	8,992	9.5
1972	76,221	6,586	8.6	1981	90,197	6,566	7.3
1973	83,915	6,028	7.2	1982	85,714	8,736	10.2
1974	104,367	4,198	4.0	1983	82,206	8,491	10.3
1975	87,409	6,145	7.0				

Source: Iron and Steel Yearbook, Statistical Office of the European Community, Luxembourg, issues 1976–84.

None the less, the apparent 'success' of the first VRAs moved the steel-making firms in the European Community to attempt export restraint negotiations of their own with Japanese producers. At first, the problematic legal status of such negotiations prevented any agreement from being concluded. In early 1971, for example, talks on a cartel-like marketing arrangement between Community (including British) steelmakers, led by the West German firm, Mannesmann, and Japanese steel exporters were cut short by threatened anti-trust prosecution by the West German Anti-trust Office (*The Economist*, 6 March 1971, p. 91).

Subsequent discussions between officials of the European Coal and Steel Community (ECSC) and Japanese producers, however, lent greater legitimacy to the proposed export restraints (*The Economist*, 20 November 1971, p. 10). On 9 December 1971, the Japanese Iron and Steel Federation announced its commitment to limit 'voluntarily' steel exports to the EC and the United Kingdom to 1.25 million tons in 1972, with quotas for shipments in 1973 and 1974 to be set by future negotiations (*Wall Street Journal*, 9 December 1971, p. 26). Still, it is noteworthy that this announcement found its way into print only on the back pages of the *Wall Street Journal*, with the Japanese alone giving notice of the agreement. The low profile of Community and British producers in the matter reflected at once the political and legal tenuousness of the export restraint agreement. Politically, the lack of a groundswell of protectionist sentiment in the Community or the United Kingdom left the agreement without camouflage as a patent accommodation of producer interests. Legally, such collusive arrangements in international trade remained questionable, leading to the obsession among all parties to project an image of the 'true voluntary spirit' of the Japanese exporters who made the announcement.

In practice, the agreements between Japanese exporters and EC and British producers appeared to have little effect on steel trade. Japanese exports in 1972 to the countries in question totalled 1.61 million tons, exceeding the agreed limit by 0.36 million tons. As noted earlier, the favorable steel market conditions of 1973 and 1974 made restraint agreements in those years superfluous. Nevertheless, the agreements set an important precedent for subsequent export restraint agreements negotiated by Community officials with Japan and other countries.

In addition to the example this agreement provided for future experimentation by policymakers, it showed clearly how one instance of export restraint led to similar arrangements elsewhere. This first 'reflex' action to ward off imports diverted by such policies established a pattern of imitation and rebounding protectionism which was to carry over into the next period of policy crisis in the steel industry. In a

sense, all subsequent uses of 'voluntary' restraint in steel trade policy were direct descendants, the 'policy children' of the first VRAs.

The most important aspect of the 1971 Japanese—EC agreement as regards the study of 'voluntary' restraint, however, lay in its exposed incentive structure. With no governmental organs yet available (or willing) to conduct negotiations, producers pursued their aims through a direct agreement with their competitors, a situation reminiscent of the International Steel Cartel of the 1920s and 1930s. While other policies of 'voluntary' restraint have been implemented or concluded by government action, this one instance revealed the pattern of interests which is served by all such agreements. In the end, 'voluntary' restraint represents a marketing arrangement designed to accommodate *producer* interests.

Summary: policy implications of the first VRAs

The importance of the early steel export restraint agreements lay not so much in their direct effects on trade and welfare (which were not large) as in their impact on policy thinking. Policies of negotiated export restraint were found to offer a viable, if imperfect, alternative to the traditional instruments of trade restriction. The policy 'learning process' of 1968—74 brought out the weaknesses of the first agreements which would need to be corrected when pressure for trade restrictions on steel surged again in 1977. The 'lessons' of the early agreements centred on the following problems.

Shortcomings in enforcement   While the first VRAs appeared to effectively reduce steel imports to the United States during a period of threatened protectionist legislation, a more systematic method of intimidation was needed in order to ensure that the exporters under 'voluntary' restraint would not 'cheat'. In other words, a consistently effective and viable threat of alternative import restraint was needed to guarantee compliance.

Shortcomings in specific product coverage   Part of the enforcement problem was actually a matter of controlling imports in specific product categories, since export limits on broadly defined steel categories merely created the opportunity to substitute more expensive products, refocusing the trade disturbance on specialty steels. A comprehensive system of minute, product-by-product coverage would be necessary to make a policy of induced export restraint work.

Precarious legal status of collusive restraint   Ironically, it is probable

that the non-binding nature of the first pair of VRAs at once doomed the agreements to ultimate failure and saved them from conviction under the anti-trust laws. To be sure, their legal status was eventually more a matter of concern among United States officials than among Community producers. Still, the delicate legal task remained for all policymakers to avoid anti-trust restrictions while effectively inducing the exporter to reduce his supply of the goods.

Policy rebound effect of diverted trade   The emergence of 'voluntary' restraint as trade policy in one country quickly led to its implementation elsewhere. Any major trade restrictions concluded outside a multilateral framework refocus exports on other, less protected markets. Policymakers in these markets then come under political pressure to devise similar trade restrictions. The 'lesson' for policymakers was to anticipate other importing areas' trade-diverting policies and to have one's own protectionist devices at the ready. The practical effect was to create a pattern of spiralling proliferation of such measures.

The introduction of 'voluntary' restraint as an instrument of trade policy thus set a precedent which sought imitation and refinement. The first, rough-hewn VRAs of 1968 led both to the more refined VRA renewals of 1972 and to the producer-negotiated restraint agreements between suppliers in Japan and the European Community of 1971. Although steel market conditions improved in 1973 and 1974, allowing policymakers to let the agreements lapse, the worldwide downturn in demand in 1975 prompted their resurgence. The proven appeal of export restraint as a remedy to steel market 'disruption' set the stage for the use of more extensive and discriminatory import restrictions in the United States and the EC during the next policy crisis in world steel trade.

Notes

[1]   Senate Bill S2537 of 16 October 1967, 90th Congress, 1st session, 1967. See Lowenfeld, 1979, p. DS473.
[2]   Preeg, 1970, pp. 103–6. See also the testimony of Ambassador Roth, chief US negotiator at the Kennedy Round, on the dangers of quotas for the Kennedy Round accomplishments in US Senate, 1967, pp. 34–7.
[3]   Lehnardt, 1969, p. 108. It is noteworthy that the Japanese Ministry of International Trade and Industry at first opposed 'voluntary' restraints, fearing the spread of such solutions to other Japanese export industries, but later acquiesced in the negotiations after being convinced of the strength of protectionist forces in the US Congress (see p. 109).
[4]   The VER agreements that resulted from these negotiations became known as the Voluntary Restraint Agreements (VRAs) of 1969–74. In this study, the term 'VRA' will refer to these specific agreements, while the generic term 'VER' will be used to describe the policy in general, as well as other such agreements.

5   See the 'Statement of Intention of the Japanese Steel Industry,' memorandum from the Japan Iron and Steel Exporters' Association to the United States Secretary of State, 23 December 1968, and the letter from the Association of the Steel Producers of the ECSC to the United States Secretary of State, 23 December 1968, reprinted in Lowenfeld, 1979, pp. 205–8.

6   The Japanese firms made sure to point out in their VRA 'statement of intention' that they did not control the entire steel export industry, but indicated that they would attempt to secure the 'co-operation' of the remaining exporters. See the 'Statement of Intention...', paragraph 2, in Lowenfeld, 1979, pp. 205–8. The EC's 'statement of intention' made no such reference and apparently did not consider the independent firms to be an export threat.

7   See the letter from the Association of Steel Producers of the ECSC to the United States Secretary of State, 2 May 1972, and the letter from the Iron and Steel Exporters' Association to the United States Secretary of State, 4 May 1972, in Lowenfeld, 1979, pp. 476–83.

8   *Consumers' Union of the United States, Inc. vs Rogers et al.*, Civil Action No. 1029–72 (DDC filed on 24 May 1972).

9   United States Court of Appeals, District Court Circuit 506F 2nd 136 (1974); *certiori* denied, 421, US 1004 (1975).

6

The Crisis Renewed: Towards Binding Voluntary Restraint

The boom in steel demand in 1973 and 1974 masked the growing divergence in competitiveness between Japanese steelmakers and their American and European counterparts. The subsequent recession in the iron and steel industry in the United States and the EC therefore hit particularly hard, as Japanese steel exporters increased their penetration of these markets to unprecedented levels. In the first two years of the stubborn trough in demand, market conditions called for cutbacks in production, plant shutdowns, layoffs and a drop in prices (see Figure 4.1). Both steel producers and organized steel workers resisted the pressures to adjust to such severe market circumstances, however, and the rapid rise in imports provided a visible political target which could act as a focus for their protectionist efforts.

The parameters of the policy response in the EC and the United States to the call for protection were shaped by three essential factors: (1) the continued restraints on the use of traditional instruments of trade restriction; (2) the experience already logged by policymakers with 'voluntary' restraint since 1968; and (3) the development of greatly increased government intervention in trade matters in the Community and the United States. The political volatility of the steel industry's adjustment pressures, in combination with these factors, created a highly communicable form of protectionism which was spread by reactive and pre-emptive policy measures. As the policy crisis deepened, the import restrictions tended increasingly to induce binding, discriminatory export restraint.

Recession, competition and the import problem

Primary source of the trade disturbance
The recession that gripped the industrialized world in the wake of the OPEC price shock adversely affected the steel industries of the United

States, the EC and Japan. Capacity utilization in all three areas dropped precipitously in 1975 and would not begin to improve until 1978. Spot prices for imported steel in the United States, a good indicator of conditions on the world steel market in general, showed an even more dramatic decline (Figure 4.1c). In addition to the overall decline in steel demand during this period, however, the shift in cost competitiveness on world markets towards Japan had become increasingly evident. The tremendous surge in steel demand in 1973 and 1974 had therefore left producers in the United States and the EC doubly unprepared for the severe shock that followed. Not only had demand declined rapidly and unexpectedly (part of a secular trend that most steel producers failed to recognize), but the world market had become much more competitive, leaving no room for marginal producers. Under these market conditions, cost-competitive Japanese exports sparked protectionist sentiment in the most exposed market, the United States.

The most direct explanation of the relentless rise in Japan's penetration of world steel markets can be found in a comparison of her cost efficiency in steel production with that of her major rivals. The major source of shifting competitive advantage has been the difference in labor-cost efficiency (FTC, 1977, p. 522). Trends in the productivity of labor and wage rates among world producers had given Japan a clear advantage in this regard. In 1978, estimates in employment costs per ton of steel shipped revealed this competitive gap: $71.46 for Japan, $96.21 for the United Kingdom, $107.35 for West Germany and $114.10 for the United States. In general, the Japanese advantage in production efficiency can be attributed largely to consistently shrewd investment decisions and technological adaptation geared to the rigors of world export market standards. Large-scale, modern, integrated plants incorporate the latest production technologies and were built prior to the recent surge in construction-cost inflation. The location of large plants near deepwater ports and the general export-market orientation of the Japanese steel industry facilitated the inroads they made in steel markets abroad. By 1976, 44 per cent of Japanese crude steel production was being exported (Walter, 1979, pp. 176–8).

The aggressive competitiveness of Japanese exporters in both the American and Community steel import markets caused them to be regarded as the 'disrupters' of steelmarket conditions, but it would be wrong to give the Japanese alone too much credit for the situation. Market opportunities exploited by them were to a certain extent created by the deterioration of competitiveness on the part of the steel industries in the United States and the Community. In particular, trends in wages and productivity among steelworkers, the rate of technological adaptation and governmental involvement in the indus-

try suggested that firms in these countries may have forfeited a considerable portion of their markets through public and private mismanagement.

Roots of dirigiste trade policy in the European Community

When the world steel market descended into an extended recession in 1975, the EC was already burdened with considerable over-capacity in steel production. Beginning in 1968, a number of steelmakers in the Community embarked on programs to enlarge their crude steel production capacity. Real annual investment expenditure in the steel industry of the European members of the Organization for Economic Co-operation and Development (OECD) rose nearly fourfold from 1968–75, from $1.2 billion to $4.6 billion (OECD, 1977, p. 45). Much of the financing for these expansion programs was provided by government, especially those of Britain, France and Italy, which had undertaken programs of nationalization (Walter, 1979, pp. 171–2).

The EC's steel industry was therefore extremely vulnerable to the market decline in 1975 after the boom conditions of 1974. In that one-year period, average steel plant capacity utilization fell from 85 per cent to 65 per cent and average steel prices plunged 35 per cent as firms in the Community sought to maintain market shares through aggressive price-cutting. In the meantime, Community export market shares had been eroded by Japanese penetration of world markets, especially in the United States. Increased Japanese penetration of Community steel markets, while still low in absolute terms, confronted a political situation which had become very sensitive to additional market 'disruption' (see Table 5.6). Thus, the convergence of improved Japanese competitiveness, cyclical market conditions and misdirected investments set the stage for a protectionist surge among steel producers in the Community.

Protectionist sentiment in the EC followed channels defined by the structure of policymaking and the provisions for the laws governing the ECSC. In March 1975, Jacques Ferry, President of the French Steel Federation, requested that a state of 'manifest crisis' be declared in the Community's steel industry, under the provisions of Articles 58 and 61 of the Treaty of Paris (*New York Times*, 29 March 1975, p. 28). Such a declaration would have allowed the European Commission to enforce production quotas, minimum prices and quantitative import controls (See Lowenfeld, 1979, p. 278).

In spite of the rapid deterioration of steel market conditions, the European Commission was not prepared at that point to pursue such drastic measures. Both internal and external policy considerations contributed to this decision. The creation of a crisis cartel, as envisaged by a 'manifest crisis' declaration, would have generated bickering over

the distribution of production cuts, as well as outright opposition among some firms. In an industry composed of firms of varying size using productive capacity of various vintages and levels of technology, some steelmakers were in a better position than others to cope with the prevailing circumstances. In particular, many efficient, independent firms in the Brescia district of northern Italy and in other Community countries would have been difficult to control in a cartel designed to save the less efficient firms.

The 'manifest crisis' measures would also have presented difficulties in trade policy. In this regard, the European Commission's attitude was similar to that of United States trade officials faced with- the prospect of steel import quota legislation in 1968. In so far as the problems of the steel industry in the Community were caused primarily by general market conditions and not by imports *per se*, a valid application of Article XIX of the GATT (the 'escape clause') was dubious; and the use of quotas otherwise would be in direct violation of Article XI. In addition, an overt protectionist policy in the Community was likely to provoke the United States to take similar action, damaging the Community's export market opportunities. It was important to the European Commission, therefore, not to make the first move towards comprehensive steel import controls, lest policy reaction in the United States inflict even more damage on beleaguered Community producers.

Finally, the Commission evidently felt that the situation in 1975 could be met with milder, stopgap measures. Instead of a crisis cartel, it introduced a voluntary 'forward programme,' based on Article 46 (2) of the Treaty of Paris, which called for self-discipline and careful planning of investment and production decisions. In December 1975, the Commission announced minimum 'reference' prices on certain steel products, which were also voluntary. Imports were not subjected to additional direct restraints, but only to monitoring (OJEC, no. L7, 14 January, 1976).

'Voluntary' export restraint, however, also appeared to be a safe alternative to unilateral quotas. In November 1975, the European Commission concluded its first VER agreement with Japanese steel exporters in Mexico City. Japanese producers were required by the VER to limit their deliveries to the Community to 1.2 million metric tonnes in 1976. The previous threat of direct import restraints, reinforced by the growing intervention of the Commission in the European steel industry, provided an effective means of import intimidation and, in 1976, Japanese steel deliveries to the Community dropped by 25.5 per cent from the previous year's deliveries (Walter, 1979, p. 172).

In the meantime, the Commission had export plans of its own. Under conditions of continuing slack demand in domestic markets,

EC steel output increasingly became geared towards export markets. Industry Commissioner Davignon outlined his strategy of restoring health to the EC steel industry in the following manner:

> We must first get a grip on our own steel market and then, by being suitably competitive, run a significant export balance, direct and indirect ... the European steel industry cannot look toward organisation of its internal market as a means offsetting low competitiveness, for it must continue to export. (Davignon, 1979, p. 27)

One goal was apparently to recapture market shares in the United States lost to Japaneses producers (Walter, 1979, pp. 172–4). In view of the vigorous enforcement of countervailing duty and antidumping laws in the United States, the nature of steel export promotion in the Community remained secret. The EC's plan to recapture market shares in the United States, in any case, implied an aggressive export promotion strategy that would, in itself, fuel protectionist fires.

The first stage of widespread trade intervention in the EC can be summarized as follows. The trade-related disturbance of 1975 in the Community steel market emanated from increasingly cost-efficient Japanese producers and was compounded by a cyclical trough in demand. The vulnerability of Community firms resulted both from the technological ferment in the industry – which Japanese producers were exploiting more than their European counterparts – and from investments in excess steel capacity in the Community. One must, in turn, note the apparent impact of increased government involvement in steelmaking on the quality of entrepreneurial judgment which stood behind these investment decisions. Visible channels of additional government intervention in trade also helped to increase the reaction threshold level of firms to their declining competitive position. Market-driven adjustment would have required the phasing out of steelmaking capacity rendered uncompetitive by evolving world market standards, with an accompanying re-deployment of labor to other industries. The private cost of such measures to steelmakers and the disruption of widespread layoffs would in any case have been painful to absorb, indicating a structural tendency in the steel industry to delay adjustment. The scope of national government involvement in Community steel industries, however, had already triggered a process of internalized protectionist expectations, which moved firms further away from market adjustment and towards a plea for government intervention in trade.

At this point, however, protectionist sentiment in the EC did not reach the requisite threshold level to force the implementation of full-blown direct import restrictions. The European Commission,

mindful of the probability of retaliatory protectionism in the United States, chose to introduce a set of milder internal market measures to deal with the industry's problems, relying upon a broadly defined VER with Japan to control import disruption. The reliance on voluntary reference prices and export restraint pointed to the belief among policymakers that the crisis would be short, since such measures would inevitably be eroded by internal price-cutting and import penetration if depressed conditions continued. Clearly, the extent and severity of the recession had not yet become evident and therefore had not created the protectionist sentiment among policymakers which induces direct measures against imports.

Yet the most important implications of this episode of the steel policy crisis lay in the consolidation of policymaking power by the European Commission. The creation of a *dirigiste* policymaking structure changed the nature of the open political process that normally balances protectionist and free trade influences. The power of the Commission to implement crisis measures effectively 'short-circuited' this process by removing the policymaking decision from the dictates and scrutiny of a balanced political consensus. In this manner, a mechanism was established to intervene in steel industry trade by fiat, and an official framework for negotiating VERs with 'disruptive' foreign suppliers was created. While such government intervention in the steel industry was not entirely new to the member countries of the ECSC (see Dale, 1980, pp. 146 and 152-3) its increasing presence during the deepening steel recession was to breed yet more protectionism in the United States.

Vulnerability of the American steel industry

In the United States, favorable demand conditions, buoyed by the devaluation of the dollar, had silenced protectionist sentiment in the steel industry from 1973-4. A new downturn in world demand in 1975, however, led once again to increased import penetration in the open American market. Imports as a percentage of apparent supply increased to 14.1 per cent in 1976 and 1977, before peaking at 18 per cent in the following year (Figure 4.1). The surge in import penetration was accompanied by a protectionist fervor much stronger than that of 1967-8. The highly visible trade disturbance, triggered by the new cyclical demand trough, had three components: (1) the secular trend in the competitiveness of Japanese firms and of other, smaller steel producers in newly industrializing countries; (2) steel exports from Japan apparently deflected towards the United States by the Mexico City VER agreement between Japan and the EC (see Figure 6.1); and (3) the intensified efforts of Community producers to regain market shares in the United States.

GRAPH

116 POLITICS VS ECONOMICS IN WORLD STEEL TRADE

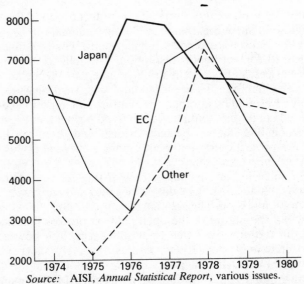

Source: AISI, *Annual Statistical Report*, various issues.

Figure 6.1 *US steel imports by region of origin, 1974–80 (thousand tons).*

Yet, as in the case of the EC's steel industry, the sensitivity to imports was aggravated by the vulnerability of the American steel producers. Some of their vulnerability can be traced to independent changes in the structure of the world market, such as the lowering of natural trade barriers (the general decline in shipping costs and the opening of the St Lawrence Seaway, for example), the market entry of new steel producers from industrializing countries, and the increase in American prices for several steelmaking inputs, such as iron ore and coking coal, relative to world prices. Still other elements of lost competitiveness can be traced, to market and entrepreneurial lassitude. Oligopolistic pricing practices, marked by a failure to recognize the need for cost competitiveness, poor performance in technological adaptation, and a growing wage–productivity gap all contributed to the American steel industry's deteriorating market condition, as described in Chapter 4.

In addition, protection from imports under the VRAs had the effect of diminishing the incentive to invest in new equipment and technologies. Ironically, the ostensible purpose of the 'voluntary' restraint agreements of 1968 and 1972 had been to allow American firms added time to close the capital expenditure gap that had contributed to their competitive decline in the 1960s. Normally, the existence of such a gap under competitive conditions would induce firms to adjust by making

the appropriate expenditures without government intervention. Import penetration during this period in the United States market showed that even an oligopolistic domestic market structure is not immune from such competitive pressures in the context of world trade. The VRAs, however, removed this incentive structure by means of a collusive agreement to reduce competition. Even though the VRAs were supposed to facilitate the required adjustment, capital expenditures among American steel firms actually declined during the VRA years (see Table 6.1). One can regard such entrepreneurial behavior as rational only under the assumption that the VRAs had established an incentive structure which caused steel firms to expect the continuation of trade restrictions in times of demand troughs.

In short, market-driven adjustment to the new competitive conditions was not followed, which meant that the severity of the industry's vulnerability increased inexorably towards a crisis level. The experience of the industry with government intervention in trade suggests that steel producers internalized such import protection into their decision-making process, thereby delaying adjustment, in anticipation of eventual government action. In this manner, the normal mode of rational entrepreneurial judgment was distorted and market channels of adjustment avoided, creating a vicious circle of vulnerability and inaction.

Table 6.1 *Capital expenditures of steel industries in selected major steel-producing countries (million $)*[a]

Year	United States [b]	European Community [c]	United Kingdom	Canada	Japan
1965	1,823	932	139	141	510
1966	1,953	848	117	187	540
1967	2,146	730	136	114	843
1968	2,307	802	119	61	1,167
1969	2,047	1,005	102	95	1,494
1970	1,736	1,615	191	193	1,889
1971	1,425	2,310	414	236	2,607
1972	1,174	2,810	411	209	2,443
1973	1,400	3,033	401	215	2,039
1974	2,104	2,850 [d]	400 [d]	300 [d]	2,700 [d]

Source: Steel Industry Economics and Federal Income Tax Policy, (Washington: American Iron and Steel Institute, June 1975) p. 52.
[a] At official exchange rates.
[b] Includes non-steel-producing activities of steel companies.
[c] The European Community here refers to the original six member countries.
[d] Estimated.

A consideration of the vulnerability of American steel producers points logically to the motives behind the ensuing protectionist sentiment. In essence, the producers' advocacy of protection was based on their desire to restore the oligopolistic market structure which was disrupted by import penetration. Steelworkers, on the other hand, had a wage–productivity gap to protect. It is noteworthy that the political economy of protectionism in this instance was shaped largely by an imperfect product-market structure, which strongly aligned both capital (owners of steelmaking capacity) and organized labor in favor of import controls.

Elements of vulnerability also shed light upon the specific call among protectionist forces for quantitative restrictions. The widening gap in competitiveness between American and Japanese producers, combined with the unpredictable amplitude of the cyclical trough, made quantitative restrictions the only reliable forms of relief from market disruption due to imports. Emergency tariff measures at politically defensible levels, for example, would be unable to stem the expected continuing tide of imports in a volatile market subjected to severe price-cutting.

The concerted efforts of protectionist forces therefore sought the passage of import quota legislation against steel imports. Nine such bills were introduced in 1977,[1] a year which can be characterized fairly as one of protectionist panic in the American steel industry. The strategy of the American steel producers and steelworkers was to: (1) mobilize the 'steel caucus' on Capitol Hill, numbering 120 in the House of Representatives and twenty in the Senate (Lowenfeld, 1979, p. 258) to press for stringent import quotas; (2) simultaneously seek import relief through antidumping laws, thereby increasing pressure on the executive branch of the government for action; and (3) appeal for public support in order to create a 'grass roots' protectionist mood.

The protectionist strategy of the steel industry in 1977 paralleled its plan of 1967 but its adroit use of mass communications' media created a much stronger protectionist mood this time around. From mid-September 1977, when the United States Steel Corporation filed a dumping complaint against Japanese producers (*New York Times*, 10 September 1977, p. 31) and several import quota bills were introduced in Congress, until the end of the year, news of the American industry's troubles and the impact of (mostly Japanese) disruption filled the newspapers almost daily. News of import-induced layoffs, estimated to number 20,000 by the end of 1977 (*The Economist*, 31 December 1977, p. 79) was highlighted in the press and on television. The Bethlehem Steel Company even took out an advertisement in the *New York Times* urging a government crackdown on 'unfair' imports (*New York Times*, 14 September 1977, p. 21). Given the massive protection-

ist pressures being applied elsewhere in the news media and through lobbying groups, that particular steel firm evidently decided that the resources devoted to protectionist advertising would earn a greater return, in terms of government-induced import relief, than they would in traditional uses to improve the efficiency of the firm.

The policy crisis which emerged in 1977 was also a reprise, in many respects, of the 1968 protectionist surge. The executive branch under President Carter wanted to avoid, if at all possible, the implementation of import quotas, since this would have violated the GATT and could have led to an all-out trade war, spreading to other traded goods. An increase in tariffs, fixed by trade agreements, was also out of the question for practical reasons, and search for new ideas was begun. Thus the problem in 1977, as in 1968, was to find a viable alternative protectionist policy.

Antidumping and trigger prices in the United States

In the meantime, American steel producers were seeking relief from import competition through antidumping laws. In principle, the use of antidumping measures was also objectionable to the Carter Administration, since it could have led to retaliation in kind, especially on the part of the EC. The basic risk governments incur by allowing the use of antidumping laws as an instrument of import control is that they deprive the government of full rein over import restrictions. Throughout the first part of 1977, protectionist sentiment was focused on the apparent source of market disruption, Japan, and the record of antidumping investigations reflected this attitude. The surge in Japanese imports had already resulted in a dumping complaint by the Gilmore Steel Company, filed in March 1977 (*Federal Register* 42:16,883, 1977) and led to further rumblings of possible discriminatory import restraints against Japan that year (*New York Times*, 20 September 1977, p. 1). The United States Steel Corporation filed its own suit against Japanese producers in September, timed to coincide with the introduction of quota legislation in Congress.

Dumping enforcement: the impact of the Gilmore case
Antidumping enforcement tends to induce export restraint among the lowest-cost, 'disruptive' foreign producers, while higher-cost imports, even if dumped, often escape enforcement. The initial effect of the antidumping enforcement in 1977 was to deter imports from protectionist-targeted Japanese producers in exactly this manner. Although the bureaucratic process was laboriously slow, the antidumping investigations had an intimidating effect on imports. As the

investigations proceeded during 1977, Japanese deliveries to the United States of those products subject to dumping complaints dropped precipitously.

Much of the induced export restraint by Japanese producers can be traced to the fact that the Gilmore case marked the first use of cost-of-production criteria, established in the Trade Act of 1974, in a steel dumping investigation (see Rodriguez, 1979, pp. 186–201). Under this type of antidumping probe, the foreign firm must demonstrate that the prices it charged enabled it to recover full costs of production, including a 10 per cent mark-up for fixed costs, plus a 'normal' rate of return set at 8 per cent. The Trade Act of 1974 stipulated that cost-of-production criteria be used in dumping investigations if less that 10 per cent of the foreign firm's total sales during the investigation period had taken place in its home market at prices above the cost-of-production (US Code 19, 164b). The protectionist effect of the cost-of-production criteria are especially clear in the case of steel. Since the steel industry is subject to high fixed cost, steel prices will often drop below the *average* cost of production during cyclical market downturns, as firms maximize profits (or minimize losses) by setting marginal revenue equal to *marginal* (not average) cost. The new antidumping criteria effectively prohibited foreign firms from marginal cost pricing during recessions, a standard not applied to domestic firms. Indeed, the criteria actually implied that foreign firms must *raise* their price during a recession, since unit fixed costs would typically rise as capacity utilization declined. Adding an 8 per cent profit margin to prices during a business downturn, in any case, would not normally represent sound business practice in the steel industry. In addition, the cost data used to construct the 'fair' price would be difficult to gather and could be subject to manipulation.

The investigation therefore required the gathering of detailed production cost data, but Japanese firms were loth to submit sensitive information on their costs for fear of unauthorized disclosure to their competitors.[2] As a result, in the early stages of the Treasury Department's probe, investigators were dependent on data supplied by the Gilmore Corporation itself. Both the decision to initiate the investigation and the tentative determination of sales at less than fair value, leading to withholding of appraisement, were based largely on Japanese production cost estimates provided by the Gilmore Corporation, since Japanese producers balked at delivering such confidential data to United States government officials. A preliminary weighted average dumping margin on Japanese steel imports was set at 32 per cent (*Federal Register* 42:22937, 1977).

In order to appreciate the trade impact of the Gilmore case, however, its progress and the Japanese reaction to it must be examined in its

political context. The inquiry was conducted as protectionist fervor was reaching a peak, with a preliminary determination of dumping set by law to be given by 6 October 1977. The Treasury Department found itself under considerable political pressure to secure a 'conviction,' since American steel industry lobbyists and American legislators sympathetic to the industry's plight had based their call for import restrictions on dumping or other unfair trade allegations. In addition, the Carter Administration feared that an exoneration of Japanese producers from dumping charges would only increase pressure for quota legislation. Finally, the preliminary determination was not final and could be reversed later as better information was acquired for the investigation. Under these circumstances, the pressure for a 'tentative' determination of dumping was intense.

For the exporting steel firms, however, there was nothing 'tentative' about the protectionist attitude in the United States and its penetration into dumping enforcement. These steel producers faced a dumping inquiry based on confusing and unpredictable standards of 'fair value,' with most of the information used in the investigation being supplied by the complainant. Even if the dumping charges were dropped, Japanese steel exporters faced the prospect of stringent and possibly discriminatory import quotas. As the investigation progressed, word reportedly spread among Japanese producers to partially withdraw quietly from American markets in order to 'avoid trouble' (Kiers, 1980, p. 60). The ensuing sharp decline in Japanese shares in the American market showed that the enforcement of antidumping laws under protectionist influence worked much better, and with greater severity, to deter imports than any other form of 'voluntary' restraint (Jones, 1981, Chapter 6).

The final determination in the Gilmore case serves only as an ironic footnote to the story, since its protectionist effect was realized at earlier stages of the investigation. After the preliminary determination of dumping was issued, the Japanese steel firms under investigation finally decided that their unwillingness to disclose production cost information was doing them more harm than good, so they agreed to disclose the data to Treasury Department officials (Rodriguez, 1979, pp. 189–90). Using this new data, the Treasury Department determined that enough sales (10 per cent) above cost-of-production during the investigation period had taken place in the home market to set aside the constructed-cost criteria, as required by law (*Department of Treasury News*, 9 January 1978). New dumping margins, now based on the difference between domestic and export prices, were set at 5.4 to 18.5 per cent, depending on the particular product, and were later revised downwards to between 4 and 13 per cent (ITC, 1978). The final weighted average dumping margin of 7.9 per cent was less than one-

quarter of the figure alleged in the preliminary determination. In the context of the protracted dumping investigation, however, the 'tentative,' but official, allegation proved to induce more effective restraint on trade than the dumping duty itself.

Meanwhile, the tentative 'conviction' of Japanese steel exporters in October 1977 has raised hopes in the Carter Administration that the policy crisis could be resolved through existing trade laws. Shortly after the Treasury Department's preliminary finding was announced, President Carter promised to increase antidumping enforcement as a means of controlling steel imports (New York Times, 14 October 1977, p. 1). In spite of the problematic nature of the use of such measures as an instrument of trade policy, trade officials evidently believed (or hoped) that American steel producers would continue to direct their dumping complaints against Japanese exporters and not against exporters in the EC, whose Commission was in a protectionist mood and could easily have introduced retaliatory measures.

By late 1977, however, the protectionist and antidumping campaign against Japanese imports had shifted the source of trade disruption to producers in the EC. The intimidating effects of these developments on Japanese exporters opened market opportunities in the United States which Community steel suppliers pursued aggressively (see Figure 6.1). Encouraged by the President's promised antidumping vigilance, American steel firms filed three new dumping complaints against 'disrupters' in the Community.[3] This turn of events made clear to the Carter Administration that traditional antidumping enforcement had deprived the government of effective control over trade policy. With a renewed sense of urgency, US trade officials set about the development of a policy alternative which would simultaneously defuse protectionist sentiment, return control over trade policy to the government and satisfy all the legal and international constraints which so severely circumscribed their field of action.

Advent of the trigger price mechanism
While traditional antidumping measures were proving unsatisfactory as an instrument of trade policy, the United States government was still under heavy pressure to take some definitive action to curb imports. The protectionist campaign launched by the American steel industry threatened to lead to import quota legislation, unless some intermediate measures were implemented.

Japanese and EC exporters were equally aware of the protectionist mood in the United States, and both attempted some preventive trade diplomacy of their own. Their suggestion was to renew some sort of 'voluntary' export restraint agreement with the United States. Japan was the first to offer such an arrangement, as she was the initial target

of protectionist and antidumping action (see *New York Times*, 25 May 1977, p. IV–8). The Community came up with a similar plan as antidumping enforcement shifted in its direction in late 1977 (*New York Times*, 11 October 1977, p. 51). The fact that producers offered to re-introduce collusive, trade-restricting arrangements once again reveals the interest primarily served by such export controls.

The policy experience of the earlier VRAs, however, had soured American steel producers on the idea of non-binding export restraint. A high-ranking executive of the United States Steel Corporation had openly stated that the VRAs 'did not help much,' and industry opposition made a revival of such arrangements politically infeasible (Lowenfeld, 1979, p. 252). The major drawback to 'voluntary' restraint, from the import-competing firms' perspective, was that it was not mandatory. At the same time, the earlier VRA lawsuit reminded government officials that any kind of binding export restraint – indeed, anything following a formal VRA formula – still carried the stigma of a violation of the spirit of anti-trust principles. Yet policymakers were still attracted to the concept of 'voluntary' restraint by exporters, since such an arrangement allowed the importing country to avoid GATT restrictions on import barriers. What trade officials sought, then, was a policy device which would reliably induce export restraint across the wide spectrum of steel products and at the same time remain within anti-trust laws.

The trigger price mechanism (TPM) was expressly designed to fulfil these requirements. Announced in December 1977 and implemented in March 1978, it induced 'voluntary' restraint on the part of exporters by establishing a system of dumping reference prices, sales below which resulted in an investigation. The trigger price mechanism therefore sought to create the same intimidating effects as those of a dumping investigation, except that it allowed the government to 'pull the strings.' It was thus necessary to make the implementation of the TPM conditional upon American firms' withdrawing antidumping complaints. American steelmakers, glad to see the government finally take some action against imports, agreed (for the time being) to drop their cases against the EC. Aside from this policy advantage, the TPM covered all steel categories, with a comprehensive set of reference prices, ensuring product-by-product compliance. Finally, the trigger price mechanism was legal by GATT and anti-trust standards, in that it operated through traditional antidumping laws.

The flexibility that the TPM afforded American trade officials also allowed them to tilt the policy towards European Community interests (see *New York Times*, 14 November 1977, p. 53). The thrust of the new policy was clearly discriminatory and trade-diverting, much more so than traditional antidumping laws, since it sought to restrict imports

from low-cost dumpers only. Under this system, the largest low-cost producers in Japan were prevented from selling at below their calculated average production cost (on which the trigger prices were based), while the generally higher-priced producers in the Community were given a free dumping margin with which to compete in the United States. It is therefore not surprising that Jacques Ferry, head of the Community's steel producers' cartel, magnanimously declared that the Community's steel producers 'don't have any problems' with the TPM (*Wall Street Journal*, 3 October 1978, p. 2). Its effect was to allow EC steel producers to gain market shares in the United States from their Japanese rivals.

The initial trade impact of the TPM was, generally, to continue the effects of the antidumping and protectionist campaign of 1977. Japanese market shares increased slightly in the first quarter of 1978, after the threat of dumping investigations was removed and before the TPM became operational. During the rest of 1978, Japanese shares resumed their decline, although much of the decline may have been the result of the fortuitous appreciation of the Japanese yen in foreign exchange markets from late 1977 through mid-1978 and a conscious disengagement by Japanese producers from the US market to avoid further discriminatory protectionist actions (see Patrick and Sato, 1982, pp. 216–18). From its peak of 55.9 per cent of the American import market in 1976, Japan's market share had declined to 40.5 per cent in the antidumping year of 1977 and reached a low point of 30.7 per cent in 1978, the year when the TPM was implemented (see Table 5.2).

The TPM also initially accomplished its political purpose: to defuse protectionist sentiment at home and avoid retaliatory measures by the EC. Largely on the strength of the Japanese withdrawal from the American import market, total import penetration declined from a 22 per cent share of total consumption in May 1978 to a 14 per cent share in December of that year (*Washington Post*, 7 February 1979, p. D–7). The continued efficacy of the TPM as an instrument of trade policy depended, however, on its ability to control 'disruptive' imports and thereby keep protectionist pressures in check. As long as Japan remained the 'disrupter,' the TPM accomplished this political purpose. As Japan's 'disruptive' role declined and other countries' import penetration increased, a new policy crisis arose.

Summary
The United States steel industry felt the full effect of its 'crisis of adjustment' in 1976 and 1977. Its reluctance to pursue market channels of adaptation to new competitive conditions led to a plea for protection, supported by the formidable political influence of the industry.

Protectionist pressure reached the requisite threshold level for a policy of import restrictions to be implemented. The shape these restrictions took was determined by certain policy constraints and the fear of foreign retaliation, particularly from the EC.

The roots of protectionist pressure can be traced to the interrelated problems of vulnerability and trade-related disturbances. The vulnerability of American steel producers had a number of sources, including both secular trends and induced competitive decline. The isolation from and disregard for world market conditions contributed heavily to the industry's vulnerable position. In the context of an oligopolistic market structure, the American industry forfeited part of its market position through its failure to adopt new technologies in a timely fashion and through its wage structure, marked by a growing gap between productivity and wage rate increases.

Yet it was only when trade disturbances began to lower import prices that the industry's vulnerability translated into competitive decline. Increased Japanese competitiveness and later, a surge in EC exports drawn in by the withdrawal of the Japanese from the US market and encouraged by the Community's *dirigiste* policy, were the immediate causes of import 'disruption.' Import-competing steel firms in the United States, however, had always been reluctant to adjust to their deteriorating competitive position via market channels. The nature of the production process of steelmaking, which involves large facilities with limited alternative uses, made the stakes of adjustment extremely high. In addition, the expectation of government protection from imports gave firms an apparent alternative to costly market adjustment.

The rejection of market-driven adjustment by the United States steel industry led to a plea for protection, as the industry's approach to adjustment moved from the economic to the political realm of entrepreneurial action. The steel industry commanded formidable political influence, a result in part of a well-financed and well-organized lobby and in part of a public protectionist mood which was used to generate such trade policies. The political strength of protectionist sentiment was in fact clearly sufficient to force some sort of import control. The anticipated effect of threatened unilateral import quotas caused trade policy officials in the executive branch to seek intermediate measures. The 'policy crisis' of 1977 was a result of the difficulties in finding an instrument of import control that could simultaneously satisfy the existing policy constraints (GATT rules, foreign retaliation, political feasibility) and quell protectionist sentiment.

These circumstances pointed towards the ultimate implementation of a policy of induced 'voluntary' export restraint through the TPM. Pressure for quick results and the effect of anticipation of threatened

import quotas pushed trade officials towards an administrative mode of policymaking – which is an important feature of 'voluntary' restraint. Domestic legal constraints and GATT rules made it necessary that the apparent source of the trade restriction be the exporter. The prospect of foreign retaliation from the EC as well as the prominent role of Japanese producers as the disrupters (at least until late 1977) encouraged policymakers to devise discriminatory trade controls against Japan. Finally, the deteriorating competitive position of the American steel industry, exacerbated by the existence of a severe cyclical trough in demand, caused the industry specifically to demand supply-side restraints on trade as opposed to tariffs.

Policy constraints and the political environment in the United States thus led to the use of antidumping laws as the means of achieving 'voluntary' export restraint on the part of Japanese producers. Initially, the mere enforcement of these laws in the context of a surge in protectionist sentiment achieved this goal. Later, when antidumping suits were filed against producers in the EC, thereby violating the policy constraint of avoiding foreign retaliation, the executive branch introduced a variant of the antidumping instrument, the TPM. The TPM maintained the pattern of import trade diversion which had been accomplished earlier. Shares in the United States import market of the original 'disrupter,' Japan, dropped precipitously, allowing increased market penetration by Community producers. The implementation of a comprehensive policy of steel import controls managed temporarily to defuse protectionist pressure for more stringent measures, as the American industry agreed to wait and see if the new policy would work.

Policy rebound in the European Community

As United States steel producers were lobbying for stringent import quotas, their counterparts in the EC were becoming disillusioned with the ability of stopgap voluntary price and production guidelines to see the industry through the crisis. The Commission's 'forward programme' had failed to adjust the steel industry's output to continually sagging demand in Community markets. As a result, the problems of overcapacity and depressed prices remained and highly competitive firms, such as the Bresciani group in northern Italy, made life increasingly miserable for the less efficient producers. In addition, Community producers were accusing Japanese exporters of cheating on the 1975 VER (*The Economist*, 16 October 1976, p. 136). The failure of market conditions in the Community to improve, as Community officials had so sorely hoped, meant that any comprehensive,

Community-wide policy to support the steel industry would need to impose more far-reaching, market-defying controls on prices and imports.

The dynamics of dirigisme

In this manner, the peculiar dynamics of *dirigisme* led to the EC's increased intervention in the steel industry. The problem faced by Community policymakers was that they had introduced measures designed to induce cartel-like co-operation and restraint, but had not bothered to include an enforcement mechanism of (1) mandatory production and/or price controls and (2) restrictions on market-disruptive imports. As steel market conditions worsened, the Community moved to tighten its crisis plan. In October 1976, it organized a Community-wide steel producers' cartel, Eurofer (*The Economist*, 16 October 1976, p. 136). In May 1977, the new Commissioner of Industrial Affairs, Viscount Etienne Davignon, invoked, for the first time, Article 61 of the Treaty of Paris and set a mandatory minimum price for concrete reinforcing bars, a steel product particularly susceptible to price-cutting. He also introduced reference prices for six other steel products. In order to prevent imports from undermining these measures, his plan: (1) prohibited producers in the EC from aligning their prices with import prices; (2) introduced an intensified system of import monitoring, which included mandatory licensing; and (3) suggested the implementation of uniform antidumping measures (OJEC, no. 114, 5 May 1977, p. 1). These features of import regulation in the Davignon Plan were in fact precursors to actual import restraints and were intended as a signal that any market disruption by imports would meet with restrictions in the future.

The long-range objective of the Davignon Plan was ostensibly to make the EC more competitive on world markets. The usual market channels of adjustment, which would include writing down the book value of excess steel capacity, closing a number of plants and laying off redundant workers, was rejected as too painful politically to be borne by governments in the Community. Instead, a directed policy of 'restructuring' and 'rationalization' was designed to nurse the industry back to competitive health.

Under such a plan to replace market forces with government directives, however, it was difficult to prevent a supposedly comprehensive policy from taking on a life of its own. After the measures of the first forward program failed to reverse the deteriorating market conditions, the natural bureaucratic tendency was to try to finish the incomplete job of intervention. Thus, the Davignon Plan was born and the trend towards yet more intervention continued. In this manner, protectionism was generated within the policymaking organ and trade policy in

general came to be conducted as a form of administrative crisis management.

Protecting Eurofer from imports

The reference prices established by the Davignon Plan in May 1977 apparently enjoyed general compliance in the first few months of operation, in spite of rumors of under-the-table price-cutting (see Jones, 1979, p. 150). The Commission in fact felt confident enough about their effectiveness to raise the reference prices 2.5 per cent to 14 per cent in July of that year (OJEC, no. C174, 22 July 1977, p. 2). Yet the world export market in steel was still marked by heavy price-cutting and imports into the EC were preventing producers there from expanding their capacity utilization. One factor behind this was the growing protectionist trend in the United States, where pressure was mounting against exports from the EC. Spain and some East European countries were also disrupting the highly controlled Community market.

The European Commission desired a comprehensive set of import controls to protect its steel cartel. Policymakers handled this aspect of sectoral planning for steel very delicately, in view of the potential protectionist backlash from the United States. Since the Community was still a net exporter of steel and was trying to recapture market shares in the United States in 1977, any precipitous action on the part of the Community to restrict imports (and thereby to divert more world exports towards the United States) would have risked an angry American reaction, possibly leading to discriminatory import restrictions against the Community.

It is therefore not surprising that the European Commission announced the implementation of its basic price system (BPS) in December 1977, shortly after United States Treasury officials unveiled their trigger price mechanism. By waiting for the United States to take the first major protectionist step, the Community avoided retaliatory action. In addition, the Community's policymakers were able to appropriate the main features of the TPM for their own purposes. Given the severe policy constraints faced by both the United States and the Community there was a general dearth of ideas for politically and legally viable protectionist devices to deal with the steel industry's problems. The introduction of a new, apparently workable policy instrument in the United States was thus quickly copied in the Community.

As a policy instrument to induce 'voluntary' restraint, the BPS represented a further refinement of the TPM. The BPS, like the TPM, established 'fair value' import prices, calculated on Japanese production cost data, for all traded carbon steel products. Sales at below the

basic price resulted in an immediate antidumping investigation. A preliminary determination of dumping was declared within a few weeks of a violation and provisional antidumping duties were assessed immediately thereafter (see van Bael, 1978). The BPS thereby established *de facto* minimum import prices. At the same time, the policy of the European Commission was to offer to withdraw the BPS and the antidumping investigation for those steel suppliers who agreed to conclude a 'voluntary' restraint agreement with the Community. For producers in those exporting countries where anti-trust laws allowed such arrangements, the VER was difficult to resist. Under the VER, the supplier was allowed to sell to the EC at below the reference prices, with producers operating in the Community prohibited from aligning their prices with import prices (Lowenfeld, 1979, p. 290). This virtually guaranteed the VER participants market shares in an otherwise highly protected import market. As noted earlier, such arrangements also served producer interests by allowing scarcity rents to accrue to suppliers. In view of the threat of alternative antidumping action (which in the EC entailed both an immediate fine and the possiblity of a cost-of-production investigation), the offer of guaranteed access to the European market was especially appealing.

Under the implementation of the BPS on 1 January 1978, a large number of antidumping complaints were filed by the European Commission, and it took only a few months for the Community's principal suppliers of steel to negotiate detailed, product-by-product VERs. The most amicable 'voluntary' restraint settlement came with the EFTA countries (Austria, Finland, Norway, Sweden, Portugal and Switzerland), which did not receive antidumping fines. It took the persuasive power of preliminary antidumping fines to bring steel producers in other countries to the VER negotiating table. Japan, the Republic of South Africa, Czechoslovakia and Spain all concluded VERs within a few months of the imposition of initial antidumping duties. For those steel suppliers who were even more stubborn, the fines were extended. Hungary, Romania, Australia, Poland and South Korea finally yielded to the threat of definitive antidumping duties and concluded VERs as well. Although details of the VERs are not available for public scrutiny, the general format of the agreements included quantity limits by weight on each individual category of imported steel, generally set below 1976 import levels, and the respective margins by which suppliers were allowed to undersell the reference prices, introduced in the EC in 1977 (Lowenfeld, 1979, p. 292).

In so far as the BPS−VER policy sought to establish a comprehensive orderly marketing agreement among steel suppliers to the EC, it was also discriminatory and import trade-diverting in its dynamic if not its stated effects. By freezing market shares, the 'voluntary' restraint

arrangement prevented shifting patterns of competitiveness from being transformed into corresponding shifts in trade patterns and market shares. In addition, VERs inevitably excluded some firms or potential market entrants by means of rigid quotas in the exporting countries.

Yet this policy of inducing 'voluntary' restraint achieved its purpose of reducing steel imports into the EC (Table 5.7). In the broader policy context, however, such import restrictions were intended primarily to shore up the unstable workings of the Community's steel cartel. Thus, the introduction of basic import prices was accompanied by a three-step, 15 per cent increase in official Community reference prices, the extension of mandatory minimum prices to merchant bars and hot-rolled coils and tightened enforcement measures (OJEC, no. L352, 31 December 1977). The efforts to protect the Community market from import disruption also went hand-in-hand with a crackdown on Community minimum price violaters (OJEC, no. C133, 7 June 1978). With imports no longer providing competition in the depressed Community steel market, it remained to be seen whether producers could collectively restructure and rationalize the steel industry, guided only by a crisis cartel and its interventionist trappings.

Summary

The deepening involvement of the European Commission in the steel market ultimately reduces to a pattern of internal crisis management, with entrepreneurial incentives and public debate relegated to secondary roles. The failure of the original voluntary guidelines to stabilize the market left the steel industry vulnerable to the continuing general downturn in demand and the increasing penetration of imports. As the Commission attempted to tighten control on prices, the industry became still more exposed to imports. Finally, the Community created a steel producers' cartel, which would clearly have collapsed if imports had not been restricted. By increasing the import vulnerability of its internal market, the Community exacerbated the protectionist pressure from within its own policymaking structure. Since government intervention during market downturns had, by 1977, been taken for granted, the import pressure which reached the reaction threshold of firms simultaneously caused protectionist sentiment to reach a policy threshold level. Thus, in the absence of an open process of public debate on the issue of steel import restrictions, protectionist pressure bypassed the political stage of policy development.

The main policy constraints faced by the European Commission included GATT restrictions and the prospect of foreign retaliation. In this case, the GATT rules helped to determine the form taken by the Community's import restrictions (induced export restraint), while the

pure fear of retaliation, especially by the United States, determined the method (individual negotiations with exporters) and the timing (post-TPM) of the policy's implementation. In addition, the administrative powers of the European Commission diminished the importance of the reaction of steel consumers in the Community as a potential policy constraint and this increased the policymakers' ability to set severe restrictions on imports.

In this context, the EC was able to act decisively on the factors which motivated protectionist action. The pressure for quick results pointed towards the use of administrative measures: expedited antidumping enforcement was ideally suited for this purpose. The characteristic import vulnerability of the Community's steel cartel and the identifiable sources of import disruption pointed to quantitative restrictions on a country-by-country basis. The combination of basic prices, enforced as minimum import prices, and the alternative of 'voluntary' restraint agreements represented the ultimate and most restrictive use of induced 'voluntary' restraint as a means of import control.

Notes

[1]   House of Representatives Resolutions 9026, 9162, 9243, 9273, 9425, 9834, 9894, 9935 and 10113.

[2]   While disclosure rules officially protected the confidentiality of sensitive information, foreign firms could still have regarded the submission of such information as risky. The risk of disclosure appeared to increase under the terms of the Trade Agreements Act of 1979. See Patenode, 1980, p. 218.

[3]   The Georgetown Steel Company suit (against carbon steel wire rod from the United Kingdom), filed 11 November 1977; the National Steel suit (against cold-rolled and galvanized steel sheets from EC producers), filed 20 October 1977; and the Armco suit (against carbon steel plate, structural shapes, cold-rolled sheets and coils, steel wire rods, hot-rolled bars and bar shapes from the United Kingdom), filed 2 December 1977.

7

The Collapse of the TPM and the Establishment of Comprehensive Steel Pacts

The implementation of the TPM and the BPS/VER policy revealed the lengths to which policymakers in the United States and the EC were willing to go in the manipulation of trade laws for protectionist purposes. By subjecting steel imports to investigations of 'fair value,' trade authorities attempted to reduce the politically volatile issue of protectionism to a technical question of guilt under existing trade laws (see Finger *et al.*, 1982, p. 452). Aside from the advantages of a low political profile achieved with this approach, the TPM and BPS/VER policy allowed policymakers greater flexibility in targeting trade restrictions than conventional, non-discriminatory import barriers would have. Through selective enforcement, the manipulation of official 'fair value' price levels and (in the EC) negotiated export quotas, these policies appeared to give governments the power to focus trade restrictions on suppliers who were either most disruptive or least likely to retaliate. The mechanism of trade restriction in this context, unlike direct import controls, operated under a system of selective *export* restraint, achieved through a VER-cartel arrangement, *de facto* minimum import prices and/or the intimidating effects of antidumping investigations.

Previous experience with instruments of induced export restraint in the steel industry, however, had already showed the vulnerability of such policies. Under a negotiated 'voluntary' restraint agreement, there is, as in any cartel arrangement, always an incentive to 'cheat' on the quota limits. In addition, under trigger or basic price mechanisms, it quickly became apparent that low-cost suppliers would be tempted to evade the reference prices by falsifying customs documents or arranging for importing agents to make final sales at lower prices. Finally, as long as export restraint is not universal, a 'free rider'

problem arises. The artificial shortfall in supply under these policies causes the *export* price to the country of destination to rise. The premium on steel delivered to the importing country thus attracts added import supply, either from new or existing suppliers not participating in the VER or, under the TPM, from suppliers who manage to side-step enforcement of the trigger prices. In short, these factors make discriminatory systems of export restraint at best temporary stopgap measures, since market adjustment and profit incentives would inevitably tend to erode the induced premium on export prices.

The inherent weaknesses of the TPM and BPS/VER policy thus required that an upswing in world steel demand take place if a new policy crisis was to be averted. Despite a brief and modest increase in steel demand in 1978, however, the needed market boom never materialized. Enforcement problems, spurred by a renewed downturn in steel demand in 1979 and chronically depressed market conditions since that time, began to weaken both American and Community steel trade policies. The resulting major political confrontation between the United States and the EC pushed world steel trade towards a more binding and comprehensive system of cartel-like export restraint agreements.

Continued crisis and the decline of the TPM

The weakening cartel and export pressure in the Community
The combination of basic import prices and VER agreements, introduced by officials of the EC to protect the fragile producers' cartel, Eurofer, from disruptive imports, was not able to breathe life into the Community steel industry, despite its initial success at reducing imports. In the year following the implementation of the BPS/VER policy, internal demand remained in a slump and capacity utilization did not surpass 65 per cent. As layoffs and losses continued, some governments viewed the elimination of import competition as a chance to increase their involvement in their respective national steel industries in a protected market. France, through its nationalized banks, completed a virtual equity takeover of the French steel industry in October 1978 (*Wall Street Journal*, 12 October 1978, p. 16). One month later, the Belgian government announced the nationalization of its two major steel firms (*Wall Street Journal*, 27 November 1978, p. 6). These developments did not bode well for the competitiveness of the Community's steel industry, which as a result had 50 per cent of its production under government control.

Yet, despite the ability of the trade policy of the EC to eliminate

import competition, the market structure of the Community's steel industry, which included several smaller firms in private hands, still prevented the cartel from maintaining market discipline. The essential goal of Eurofer, as with any industrial cartel saddled with excess capacity, was to reduce production and raise prices to the point where the least efficient cartel member (after 'restructuring') could remain in operation. With demand still slack and the Davignon Plan's vaunted 'restructuring' proceeding at a snail's pace, however, co-operation with such goals continued to be difficult to secure among the more efficient firms. In particular, the reference prices and even mandatory minimum prices defied effective enforcement. The failure of the existing cartel measures to improve internal market conditions thus led logically to the next step in Community intervention, the declaration of 'manifest crisis' in October 1980. As noted earlier, this official state of emergency allowed the European Commission to enforce mandatory production quotas for individual firms.

The internal problems of steelmakers in the Community also worsened the international trade policy crisis. In general, the continuing slump in European steel demand, combined with the restrictions on competition within the Community, reduced market outlets for Community steel producers and created new pressures on firms to increase exports. In addition, certain provisions of the 'manifest crisis' decision of the European Commission created export pressure by simultaneously imposing two quota constraints upon steel producers. First, production quotas were set on the maximum level at which each firm could operate. At the same time, the firm could not make deliveries to its customers within the Community in excess of the amount set by its individually fixed ratio of Community deliveries to total deliveries. Thus, if a firm's exports to non-Community countries were particularly low in a given period, it would be constrained in the amount of EC deliveries it could make and might not be able to maintain output at the production quota limit.[1] This double constraint put added pressure on firms to increase exports to that they could expand their internal Community deliveries and maximize production under the quota system.[2] Since Community steel prices were set at artificially high levels, the quota system provided firms with an incentive to dump steel abroad in the traditional sense of the term, where export prices of steel from the EC would be driven down to levels much lower than internal prices.

The policies of the European Commission showed clearly that, in the absence of robust internal steel demand, European steel recovery would depend on access to export markets. It was, therefore, essential that the United States, as the largest export market in world steel trade, remain open for Community suppliers. Surges in Community exports, however, set the stage for renewed protectionist reaction in

the United States, leading to the suspension and ultimate collapse of the TPM.

TPM enforcement problems and instability
Despite the initial success of the TPM in stilling protectionist sentiment in the United States, it became difficult for this policy to remain effective after Japan ceased to be a disruptive influence in the American import market. Ideally, the TPM was designed to accommodate a rapid determination of dumping by Japanese producers, since the trigger prices were based on production costs in that country. When it became necessary to conduct a 'fair value' investigation on steel imports from other countries, the TPM proved to be an administrative nightmare. Furthermore, its ultimate inability to curb market penetration by EC exporters led to a renewed policy crisis.

From the beginning, the TPM was awkward to administer. Unlike the BPS, which established *de facto* minimum import prices enforced by summary administrative action by the European Commission, a 'fast track' investigation under the TPM required independent inquiries by the Treasury Department and the International Trade Commission (ITC), each lasting three months. Impatient steel producers in the United States were sceptical of the ability of this bureaucratic process to grant them rapid relief from import competition and repeatedly threatened to file their own dumping complaints and resume their protectionist campaign. Because of the structure of the TPM, as noted earlier, independent dumping suits would have caused the policy to collapse.

There was, therefore, considerable pressure on those administering the TPM to obtain 'convictions' in order to prove that the policy was not merely a bluff. However, trigger price violations were not observed among major suppliers to the American market, so investigators turned to the 'small fry.' In October 1978, the Treasury Department initiated antidumping investigations against carbon steel plate imported from Taiwan, Poland and Spain, which, according to Customs Service data, had entered the United States at prices below the trigger level (*Federal Register* 43:49,875, 1978). The investigation against steel imports from Spain was quickly terminated because of the small volume of imports involved. The charges of less-than-fair-value pricing against producers in Poland and Taiwan were sustained by the investigation, however, and an order to withold appraisement on further shipments from these countries was issued (*Federal Register* 44: 7,005 and 9,640, respectively, 1979). As noted in the analysis of the Gilmore case in Chapter 6, the main intimidating effect of a dumping inquiry results from the witholding of appraisement. In the end, the case against steel from Poland was terminated after the ITC declared

that the imports posed no threat of injury to American producers (*Wall Street Journal*, 14 June 1979, p. 4). The ITC finding in the case of Taiwan was positive, however, and resulted in the levying of definitive antidumping duties (ITC, 1979).

These cases point to the economic and political shortcomings of the TPM. It tended to discriminate, first of all, against low-cost suppliers. Whether or not a firm violating the trigger prices was actually guilty of dumping, it still had to undergo an antidumping investigation. In this way, genuinely low-cost producers were intimidated from supplying the market because of the uncertainty associated with the investigation. Aside from the economic objections to the TPM, it was unrealistic to assume that producers in Japan, or any country, could be regarded as the leaders in cost efficiency for any given length of time, particularly in the context of fluctuating exchange rates. In an industry such as steel, technological progress and the increasing speed of technological adaptation could give producers in relatively low-wage areas, such as the newly industrializing countries, the ability to compete effectively even with Japanese producers, albeit on a smaller scale.

By concentrating on such small-scale producers, the TPM also failed to pinpoint the major sources of import 'disruption' and therefore failed to check resurgences in protectionist sentiment. Japan, as the original target of the TPM, had drastically reduced its participation in the American market, electing to keep a low profile in order to avoid further discriminatory protectionist actions. In fact, Japanese producers apparently began to exercise self-imposed export restraint as early as 1977, and in 1979 reportedly agreed to an informal limit on total deliveries to the United States of 6 million tons annually.[3]

Since 1977, the EC, whose producers continued to face depressed demand at home and were therefore under pressure to export, had become the main source of increased import penetration (Figure 7.1). The structure of the TPM at first allowed producers exporting from the Community to compete on a large scale with American producers without violating the trigger prices, which were set in most cases at or slightly below prevailing domestic prices. Many observers of the steel industry in the EC, including American producers and some academic and government economists, charged that these producers dumped steel on markets in the United States at prices below the cost of production (Walter, 1979, p. 166; Putnam *et al.*, 1979, pp. 21−5). More importantly, the shift in the source of the trade disturbance gave rise to growing dissatisfaction among American steel producers with the TPM as an instrument of protection. Domestic producers had argued unsuccessfully that the trigger prices should have been readjusted upward in 1979 as a result of the sharp decline in the value of the

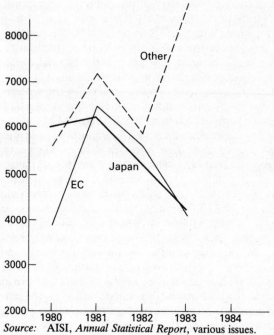

Source: AISI, *Annual Statistical Report*, various issues.
Figure 7.1   *US steel imports by region of origin, 1980–84 (thousand tons).*

Japanese yen. In the industry's view, the TPM had ceased to be an effective means of import control, and in March 1980 the United States Steel Corporation finally broke its moratorium on independent dumping complaints and filed a suit against several EC producers (*New York Times*, 21 March 1980, p. IV–3). Since this action violated the arrangement which allowed the policy to function, President Carter immediately suspended the TPM (*New York Times*, 22 March 1980, p. 1).

Whether or not Community producers were guilty of dumping under the existing criteria, the European Commission did not wish to see the matter investigated. Industry Commissioner Davignon threatened a trade war if the government pursued the investigation (*New York Times*, 10 March 1980, pp. IV–2). Thus, a policy crisis emerged again as the fragile TPM collapsed. The situation illustrated how the failure of one discriminatory measure (trigger prices) and the protectionist use of another (antidumping laws) can poison trade relations. The full story behind the ensuing policy crisis, however, stretches back to the development of far-reaching government in-

tervention in the steel industry in the United States and the EC and across the string of discriminatory trade policies used to protect the industry, beginning with the VRAs of 1968. These policies constantly refocused export pressure and accompanying protectionist action back and forth between the United States and the EC and made discriminatory devices acceptable as trade policy instruments.

The policy crisis was temporarily resolved by the withdrawal of the United States Steel antidumping suit and the reinstatement of the TPM in revised form in October 1980. The new TPM was the result of a compromise between American and Community producers' interests. A manipulation of exchange rate calculations helped to raise the trigger prices by 12 per cent, which is the main thing American producers wanted. In addition, a conditional surge provision was introduced that would step up trigger price enforcement when import penetration reached 13.7 per cent. Greater freedom to independently file antidumping suits without endangering the TPM was also granted to American producers under the surge provision of the new plan and upgraded auditing and monitoring procedures were adopted. For producers in the EC, withdrawal of the dumping suits and reinstatement of the TPM allowed renewed access to the American market, and a pre-clearance procedure was introduced that would permit genuinely low-priced suppliers to gain an exemption from the TPM upon prior application and approval (*Department of Commerce News*, 9 October 1981).

Although this patchwork of new provisions managed to salvage the TPM for the time being, the political viability of its legalistic, price-oriented framework of import controls was rapidly deteriorating. American steel firms were demanding ironclad controls on disruptive imports, preferably in the form of stringent import quotas. Firms in the EC, on the other hand, were demanding guaranteed access to the American market. While permission to supply unrestricted quantities of imports would have been preferable to them, access in the form of fixed (and of course generous) quantitative import limits would also have been consistent with the Community's *dirigiste* policy of planned 'restructuring.' It is therefore not surprising that both sets of adversaries were dissatisfied with the unpredictable, sluggish and arcane workings of the TPM, which could guarantee neither protection nor market access. The new, rejuvenated TPM was to last less than 15 months.

Advent of the steel pacts

Final collapse of the TPM
Dissatisfaction with trigger prices grew throughout 1981. This was, however, as much a symptom of the continuing weakness of the

American steel industry in the world market as it was an indication of policy shortcomings. One must therefore consider the role the TPM itself played in sustaining the vulnerability of the American steel industry to import competition.

Stripping away its idiosyncratic features, the basic purpose of trigger prices was that of any protectionist device: to insulate the domestic industry from foreign competition. In this regard, the principal economic effect of the TPM was to delay market adjustment, leaving the American industry ultimately exposed to lower-cost, more technologically advanced foreign firms. In its first year of enforcement, the intimidation of exporters by the antidumping threat of the TPM drastically reduced the market participation of Japan, which had become the lowest-cost producer and the source of the toughest competition for American firms. The reduction in competition allowed American steel producers to raise prices in 1978 by more than the industrial average and contributed to the delay in needed plant closings and modernization. The elimination of low-cost imports also allowed the United Steelworkers' union to secure a pay increase in 1980 that, given the comparative structure of steelmaking labor costs and productivity worldwide, would have been impossible in a market open to international competition (*Wall Street Journal*, 28 May 1980, p. 1). The TPM thus also helped sustain the wage-productivity gap that had been the crucial factor in reducing the American steel industry's international competitiveness throughout the 1960s and 1970s.

The deleterious economic effects of the TPM were caused by its ability to intimidate exporters with the threat of antidumping investigations. However, by 1981, the credibility of the antidumping threat appeared to be diminishing as subtle forms of non-compliance began to appear. These included hidden rebates, false price statements on customs declarations and the establishment of importing firms by foreign suppliers to resell steel at prices below the trigger levels (see *The Economist*, 28 March 1981, p. 75; *Business Week*, 13 April 1981, p. 44). At the same time, the credibility of the TPM as a protectionist device declined sharply, generating renewed protectionist sentiment among American steelmakers.

Yet the final demise of the TPM resulted from apparently open defiance of the trigger prices by large numbers of disgruntled Community exporters. Steel producers in the EC became dissatisfied with the new trigger price plan in early 1981, viewing the new trigger prices as unacceptably high. In addition, they were not receiving the hoped-for trigger price exemptions under the pre-clearance procedures outlined in the revised TPM. From August to November 1981, Community steel exporters reportedly defied the trigger prices and increased deliveries to the United States to an average of 800,000 tons per

month, up from the average 500,000 tons per month during 1980 (*The Economist*, 16 January 1982, p. 37). American producers again became impatient with the government's apparent inability to enforce effectively the TPM and announced in October 1981 that they were planning to file new antidumping suits against producers in the EC (as well as South Africa, Taiwan, Brazil, Mexico and South Korea). By failing to ward off independent antidumping complaints, the TPM had lost its usefulness to the government as a protectionist device. As was the case with the dumping suits of 1980, the new investigations ran the risk of eliciting protectionist retaliation among several major United States trading partners, especially the EC.

In an attempt to discourage the filing of the suits, the Department of Commerce threatened to suspend the TPM once more. At the same time, it initiated a domestic subsidy investigation under the countervailing duty laws against several steel exporting countries, including France and Belgium. This action was designed to establish a framework for consultations with the governments under investigation, leading possibly to a negotiated settlement. However, the prospect of an export subsidy investigation caused the European Commission to object vehemently to the Commerce Department's action (*Financial Times*, 7 November 1981, p. 2). The continued problems of managing trade relations under the TPM were thus revealed in the progressive depletion of available contingency measures to forestall independent dumping actions by American firms against exporters in major trading partner countries.

By January 1982, the efforts of the United States government to negotiate a settlement had failed and American steelmakers formally filed suits alleging dumping and subsidization, most of which were directed against producers in the EC. Thus compromised in its efforts to find an alternative solution, the Commerce Department again suspended the TPM and halted its own subsidy investigations. The antidumping actions by American steel firms repeated the pattern established in previous cases of import disruption. Their strategy again was to use complaints of unfair trade practices as a lever to get increased protection from imports, preferably in the form of import quotas. The fear of spiralling protectionist retaliation, however, had again set the stage for a new attempt at some form of negotiated 'voluntary' restraint.

The timing and circumstances of the suits afforded maximum public exposure to the American steel producers' case and put extreme pressure on the EC to negotiate a settlement. The American steel producers were in the midst of their worst slump since the 1930s; steel production would dip to 48 per cent of capacity and unemployment in the industry would reach 25 per cent by mid-1982. In addition, many

steel industry economists had already presented evidence of government subsidization in the European steel industry and a 'conviction' was regarded by many as a foregone conclusion (*The Economist*, 16 January 1982, p. 37). The EC had been able to forestall the imposition of severe trade restrictions after the previous allegations of dumping in 1980 by raising the spectre of retaliation. In 1982, however, the American steel industry was in worse straits; the complainants were more desperate. In addition, the TPM was at this point beyond repair, leaving no visible avenues for compromise policies. Finally, in the election year of 1982, unemployment in import-sensitive industries was paramount in legislators' minds, as was demonstrated by several steel import quota bills pending in Congress thoughout the year.

The antidumping and unfair trade investigations followed the same deliberate, laborious agenda as that of the Gilmore case described earlier (see Mueller and van der Ven, 1982). This time around, however, the trade litigation was specifically designed to achieve comprehensive import protection covering all carbon steel products. American steel producers were therefore angry over the ITC's preliminary determination in February 1982, which found a likelihood of injury in only thirty-eight of the ninety-two cases filed (ITC, 1982a, pp. 1–6). Since the majority of the cases had been dismissed, domestic producers could not depend on trade law enforcement itself to provide the protection it had hoped for; too many gaps in import coverage would remain. Instead it was hoping now for a determination of high subsidy margins on the remaining cases, which could then bring the EC to the bargaining table for an 'out-of-court' settlement in the form of quotas or market-sharing agreements. The incentive for a 'co-operative' settlement from the Community's point of view rested on the uncertainty of the final decisions, the intimidating effects of any preliminary duties on trade, and the overall stigma of possibly being found in violation of countervailing duty laws.

On 10 June, the Commerce Department issued a preliminary finding of subsidization on imports from seven countries in the EC, Brazil and South Africa, requiring American importers to post cash bonds equivalent to the estimated subsidy. Many American importing firms subsequently announced that they would not risk posting bonds on steel arriving from countries alleged to have high subsidy margins unless supplying mills paid the duties (*Wall Street Journal*, 14 June 1982, p. 6). The effect of the preliminary subsidization finding was thus essentially the same as that of a unilateral and discriminatory increase in tariff rates on the products in question.

The European Commission reacted by brandishing the threat of retaliation (San Jose *Mercury News*, 12 June 1982, p. 12D), but the Community's bargaining position was weakened considerably by its

inability to disprove the charge of subsidization. In an effort to fore-stall any further damage from the investigations, the EC unilaterally offered in July to restrict steel exports to the United States of those products facing trade litigation to 90 per cent of their 1981 levels in exchange for a suspension of the probe. Commerce Secretary Malcolm Baldridge rejected the proposal, however, favoring a more restrictive agreement, which he knew would be required to withstand a potential legal challenge by US steel producers, who were counting on final determinations in the cases to cut EC deliveries by more than 10 per cent. As a result, the investigation moved slowly but surely towards a major political confrontation between the United States and the European Community.

A final determination of subsidization was made by the Commerce Department on 24 August (*Federal Register*, 47:39,332–39,379, 1982). Even though most of the original subsidy margins were greatly reduced in the final determination (a repetition of the experience with antidumping margins in the Gilmore case), it was clear that the defini-tive imposition of countervailing duties would limit the Community's access to the American market in several product lines and thereby disrupt its overall restructuring plans. Furthermore, a final assessment of duties in any amount would only encourage the filing of more unfair trade practice suits against Community producers in the future. This provided an added incentive to press for an 'out-of-court' settlement.

Negotiations between the EC and the United States continued throughout the summer and early autumn, working against a deadline of 21 October, the date when countervailing duties would become final. Both sides had agreed in principle to establish a system of binding Community export quotas. For American steel producers, unilateral import quotas imposed by Congress would have been preferable, since legislators appeared ready to grant severe import limits. Dependence on this highly visible political channel for trade restrictions, however, contained all the uncertainties and political costs that accompany an open legislative debate on public policy. By maintaining protectionist pressure on Congress and seeking relief under trade laws, American producers could pursue the next-best solution: quantitative, enforce-able export limits on steel from suppliers in the Community. For the European Commission, on the other hand, 'voluntary' export limits appeared to be the best way to avoid the uncertainty of other trade restrictions, especially threatened unilateral actions by the United States Congress. Aside from producer interests, however, the govern-ments of the Community countries and the United States also pre-ferred a policy of export restraint to more traditional trade restrictions because it appeared that such a bilateral agreement could forestall a spiral of retaliation that would escalate into a trade war. Finally, none

of the parties to the negotiations ran the risk of an indictment under American anti-trust law, in contrast to previous 'voluntary' restraint arrangements, since American trade laws explicitly allowed such agreements as a means of resolving antidumping and unfair trade suits.[4]

A serious snag in the negotiations occurred, though, when producers in the Federal Republic of Germany balked at American proposals which would limit Community exports of steel tubes and pipes, of which they were the biggest supplier. In addition, West German subsidy margins were finally calculated at either miniscule or *de minimis* levels, in contrast to the larger subsidy findings for producers in France, Britain, Italy and Belgium. The EC's hopes for a settlement depended, however, on the principle of 'burden-sharing' among all its members under the export restraint agreement. The cartelized structure of the European steel industry thus once again called upon efficient firms to make sacrifices for the 'collective good' of the industry.[5]

The US–EC steel pact

West German objections drew the negotiations out until the last minute but, on 21 October 1982, an agreement on Community steel export limits to the United States was finally announced. Once again, a system of export restraint replaced the workings of trade law enforcement.[6] In exchange for a withdrawal of forty-five charges of unfair trade practices made by eight American steel producers, the EC pledged to limit exports of ten product categories of steel to market share allowances based on projected apparent consumption in the United States (see Table 7.1). The United States Customs Service would monitor imports through a system of export licensing and enforce the 'voluntary' restraint, with the power to block specified imports if export limits were violated (*Wall Street Journal*, 22 October 1982, p. 3).

The agreement took effect on 1 November 1982 and was to run until 31 December 1985, a date coinciding with the EC's official deadline for ending state subsidies (OJEC, no. L228/14, 13 August 1981). In June 1983, EC member countries also agreed to cut 26.7 million tons of steel capacity in the Community by that date. However, the continuing reluctance of Community governments to allow excess capacity to be cut casts doubt upon the prospects for reaching this goal (ITC, 1984b, p. 201). Since continued subsidization to prop up inefficient capacity would continually expose EC producers to unfair trade practice suits, the EC faced the prospect of being repeatedly forced to the bargaining table in the future just to maintain access to the American market. In addition, the temporary solution to the steel dispute between the

Table 7.1   *US–EEC steel arrangement*

EEC exports into US Market	Market share allowance (%)[a]
Hot-rolled carbon and alloy sheet and strip	6.81
Cold-rolled carbon and alloy sheet	5.11
Carbon-alloy plate	5.36
Structural	9.91
Wire rod	4.29
Hot-rolled bars	2.38
Coated sheet	3.27
Tin plate	2.20
Rails	8.90
Sheet piling	21.85
Steel pipes and tubes	5.90[b]

Source: Arrangement Concerning Trade in Certain Steel Products Between the ECSC and US, 21 October 1982, and accompanying fact sheet.
   [a] Based on projected US apparent consumption   (deliveries minus exports plus imports).
   [b] Target figure, subject to consultations and provisions of a separate protocol.

United States and the EC refocused the policy crisis on the Community's steel cartel arrangement. After the conclusion of the US–EC steel pact, the European Commission imposed fines on firms violating the production and internal delivery quotas, and introduced more stringent measures to assure compliance with internal minimum prices, including certificates of origin and a system of security deposits for internal EC deliveries (ITC, 1984b, pp. 201–4). Such measures revealed, furthermore, the conflict within the Community's own restructuring plan. Its ultimate goal was to make the EC steel industry competitive, yet its directives have prevented effective market competition.

A pattern of governmental activity emerged in the wake of the steel agreement that reflected the experience of previous instances of 'voluntary' export restraint. Aside from the increased government intervention throughout the Community implied by the export quotas, it became evident that protectionism of this sort would only breed more protectionism, both at home and abroad. Immediately following the conclusion of the steel pact, the European Commission announced new curbs on imports into the Community. This was to be accomplished both through cutbacks in existing 'voluntary' restraint arrangements with Community suppliers and the conclusion of new VERs with countries not previously covered, especially Brazil (*The Economist*, 20 November 1982, p. 70). The reason given for the new import restrictions was that export cutbacks by Community steelmakers

would require production to be diverted to internal markets. Foreign suppliers would therefore have to yield yet more of their scant Community market shares to protected Community producers. In addition, the extension of 'voluntary' restraint agreements to other countries was made necessary by the premium on steel imported into the EC, induced by the original VERs, which had attracted new sources of supply and more import 'disruption.' The new and more restrictive export restraint agreements concluded with Community suppliers, however, then diverted exports towards other available markets, including the United States, renewing protectionist sentiment there. Thus, the US–EC steel pact, instead of bringing 'order' to steel trade relations, in fact continued the protectionist cycle.

Renewed crisis and the escape clause petition

The anticipation of diverted exports hitting the American market, coupled with the demonstration effect of a successful protectionist campaign, immediately created more pressure for import controls in the United States after the steel pact was announced. With one set of disruptive imports squared away, American steel producers pressed immediately for a similar agreement with Japan and suggested that imports from other suppliers, such as Taiwan, Brazil and South Korea, would also need to be controlled (*Wall Street Journal*, 22 October 1982, p. 3). Furthermore, the absence of specialty steel from the US–EC agreement caused the American industry to press for a separate agreement on these products. Under the 1968 VRAs, specialty steel shipments to the United States surged dramatically as foreign producers shifted deliveries toward higher value products not covered in the agreements (FTC, 1977, p. 74).

The anticipated diversion of non-EC steel exports to the United States began immediately after the US–EC steel pact went into effect and intensified in 1984 (see Figure 7.1). This phenomenon became a politically sensitive issue as the American industry continued to suffer. 1982 and 1983 had represented the worst years for the US steel industry since the Second World War; capacity utilization in 1982 reached a nadir of 48.4 per cent and rose to only 56.2 per cent in 1983, despite the permanent retirement of 10 per cent of steelmaking capacity during the period (AISI, *Annual Statistical Report*, 1983, p. 8). Employment in the American steel industry fell by 31 per cent in 1982 and another 15 per cent in 1983, so that by the end of 1983, only 242,745 workers remained, compared with 512,395 in 1974. Financial losses from steel operations reached $2.5 billion during 1982 and $3 billion during 1983; when facility write-offs are included, losses reached $5 billion during each of those years (CBO, 1984, p. 29). Import penetration, in the meantime, dropped only slightly in 1983 to

20.5 per cent of the United States market, and began to rise again to its highest levels ever in 1984, reaching 33 per cent in July of that year.

In short, the US—EC steel pact had not accomplished its political goal of restoring health to the US steel industry. Despite the general economic recovery in 1983 and 1984, steel demand recovered only slightly, and the new surge in import competition contributed to keeping prices and domestic output low. The convergence of several factors led to the increase in import penetration. First of all, as noted earlier, increased steel import restrictions implemented by the EC in the wake of the steel pact tended to divert steel exports towards the United States. In addition, worldwide economic conditions made the United States an inviting market. The upturn in American steel demand, while not dramatic, was larger than in other industrialized countries, and the continued excess steel capacity on world steel markets caused non-EC steel exporters to rush towards the open US market. In this regard, several newly industrializing countries in particular faced severe pressures to increase steel exports, both to exploit economies of scale and to acquire foreign exchange to service their international debt. The United States dollar was also soaring to its highest levels since the collapse of the Bretton Woods agreement, drastically lowering the domestic price of all imports in the United States. Finally, American producers continued to operate at a competitive disadvantage on world markets, especially with regard to labor and raw materials costs.

Thus, several factors, of which import competition was only one, continued to frustrate the American industry's efforts to increase output and profits. Of these factors, however, the import issue provided the channel of most immediate relief. The industry once again mobilized a large-scale protectionist campaign. Its stated goal this time was a comprehensive global import quota limiting import penetration to 15 per cent of the domestic market. As was true in 1968 and again in 1977, the industry sought protection through various channels simultaneously. These included (1) petitioning for relief through section 201 of the trade laws, the so-called 'escape clause'; (2) the filing of dozens of unfair trade suits (antidumping and countervailing duty actions) against most major non-EC steel suppliers, and (3) the introduction of 'Fair Trade in Steel' bills in Congress, backed up by a carefully planned publicity campaign to marshall public and congressional support for protection.

As in earlier protectionist episodes, the steel industry was again using the trade laws to reinforce a protectionist campaign in Congress. The new twist this time around, however, was the duplicative effort of the escape clause petition, which requested essentially the same protection as the proposed Fair Trade in Steel legislation. Unlike the

antidumping and countervailing duty laws, which are designed specifically to protect against *unfair* trade, escape clause actions allow relief for a domestic industry suffering 'serious' injury due to increases in imports, whether fairly or unfairly traded.

The escape clause petition, which was filed jointly by the Bethlehem Steel Corporation and the United Steelworkers, may appear upon first impression to have been an inappropriate and unnecessarily risky strategy by the steel industry. A successful petition requires both an affirmative determination by the ITC, which must apply technical criteria whose interpretation is unpredictable from case to case, *and* a determination by the President that escape clause relief is in the national interest (see Adams and Dirlam, 1977, pp. 539–43). Moreover, the industry was basing its protectionist campaign on allegations of *unfair* trade by its foreign competitors; such charges are irrelevant in a section 201 petition and can only be investigated under the appropriate unfair trade statutes.

Yet these factors were more than offset by the overwhelming political logic of the section 201 petition. By filing on 24 January 1984, the industry could be assured by the investigation's statutory timetable of extensive news coverage of the issue during the height of the Presidential election campaign. The ITC was required to disclose its decision and recommendations on the case by late June; an affirmative decision by the ITC would then require the President to either accept, reject or propose alternative policies to the ITC recommendations (see Berg, 1982). Even if the ITC determination were negative, terminating the case, the investigation would still provide the steel industry with a highly visible forum for promoting its case for comprehensive import controls for the *entire* steel industry (antidumping and countervailing duty cases, in contrast, typically focus on small subsectors of the steel industry). A negative ITC finding, furthermore, would still allow the industry to turn to Congress or the President for unilateral action.

On 19 June 1984 the ITC delivered its decision, by a 3–2 vote, recommending import relief on most steel products (ITC, 1984c, p. 11). The steel industry, while generally pleased with the decision, was disappointed that the ITC had not recommended more comprehensive protection. Steelmakers had from the outset argued that the ITC should consider the steel industry as a whole in its investigation, whereby an affirmative determination would apply across-the-board to all steel products. Their fear was that protection limited to subsectors of the industry would cause foreign exporters to shift their deliveries towards the unprotected products. This was in fact the reasoning behind the industry's call for a comprehensive, product-by-product import quota. Yet the ITC had divided the larger steel industry into nine constituent industries, and had recommended relief in only five of

them, although these categories together covered approximately 70 per cent of all imports.[7] The industry was therefore hoping that the President would go beyond the ITC recommendations to impose comprehensive, global import controls.

President Reagan, for his part, was seeking a policy that would satisfy domestic steel producers while simultaneously satisfying diplomatic constraints. Despite a substantial lead in the election polls over his challenger, Senator Mondale, Mr Reagan risked losing electoral votes in key industrial states such as Pennsylvania and Ohio if he rejected protection entirely. Furthermore, aside from purely political considerations, the President had shown a basic proclivity towards protection for American industries suffering from competitive decline, as evidenced by the 1982 US–EC steel pact, by his insistence on continuing the Japanese VER agreement on automobiles through March 1984, and by his approval of escape clause relief for motorcycles and specialty steel.

Yet any new trade protection plan also had to accommodate the interests of United States trading partners in order to avoid retaliatory trade restrictions. In this regard, implementation of the ITC recommendations would be difficult since it would have to take place within the GATT framework, which would require a non-discriminatory application of escape clause import controls and expose the United States to retaliation or claims for compensation by unsatisfied exporting countries. Such a plan might also have required a renegotiation of the 1982 US–EC steel pact, causing more diplomatic headaches. Allowing the passage of import quota legislation, however, would have been even more disastrous in that it would violate GATT article X and practically guarantee retaliation.

The new VER arrangements

All signs therefore pointed to the introduction of some form of negotiated export restraint program. On 18 September, 1984, President Reagan announced that he was rejecting the ITC recommendations in favor of a system of VERs designed to limit import penetration to 18.5 per cent of the US market for steel mill products and 1.7 million tons of semi-finished steel annually. The framework of the new steel import plan included:

(1) negotiated 5-year VERs with countries whose steel exports to the United States had surged 'unfairly' in recent years, 'unfair' practices defined as dumping, subsidization and diverted trade from other importing countries that restrict access to their markets;
(2) coverage of all steel mill products with product-by-product limits based on historic patterns;

(3)   a system of licensing to enforce the VER limits;
(4)   the continued 'rigorous' enforcement of unfair trade laws, including section 301 (whereby Presidential action can be used to counteract unfair trade practices);
(5)   the possible termination of unfair trade cases in exchange for VER agreements;
(6)   the reaffirmation of existing VER agreements; and
(7)   consultations with American trading partners on ending 'trade distortive and trade restraining practices' in other markets (Reagan, 1984; USTR, 1984).

The new VER plan represented the culmination of many years of policy experience with voluntary restraint programs. Unlike earlier VER and trigger price policies, the new policy attempted to impose a nearly airtight, global system of binding, bilateral marketing arrangements. Export restraint was no longer 'voluntary,' since a licensing system (item 3) would enforce the quota limits. The new plan also adopted the European method of using trade law enforcement as leverage to pressure reluctant governments into VER agreements (items 4 and 5). Coverage on a product-by-product basis (item 2) would prevent exporters from switching their historical mix to penetrate different steel product markets.

At the same time, the new VER plan satisfied current policy constraints. Unlike the import controls as proposed in the Fair Trade in Steel Act, the negotiated export controls did not *directly* violate GATT rules, although they did contradict the spirit and principles of the GATT. More important was the fact that they kept trade peace with the EC by maintaining the 1982 steel pact (item 6). EC steel producers were especially pleased because the new trade restrictions would prevent an erosion of their US market share by unrestrained exporters.

In spite of the rigid provisions of the plan, Reagan administration officials insisted that the 18.5 per cent penetration level would only be the *expected* result of the VER arrangements, although the steel industry regarded it as a firm limit (AISI, 1984). They also went to great pains to characterize their plan as a rejection of protectionism in favor of measures necessary to counter unfair trade practices.

Certain idiosyncracies of the VER plan cast doubt upon the disavowal of protectionism, however, if one defines protectionism as the use of trade restrictions outside GATT rules. In order to bring steel import penetration down to 18.5 per cent, nearly all steel trade to the United States would have to be covered by VER agreements – whether or not it was traded unfairly and whether or not it caused injury to producers of the specific products cited in the ITC decision.

To be sure, relief measures for those limited steel products identified by the ITC would be justified under the GATT, subject to GATT procedures. In addition, official investigations had led to final determinations of antidumping and countervailing duty violations against many smaller exporters, such as Brazil, Mexico, Argentina and Spain, and most of these countries were willing to bargain for VERs in exchange for a termination of penalty duties and further investigations. Yet, while these unfair trade determinations specified import restrictions only for limited categories of steel products in each case, the VERs were to restrict *all* steel exports from the countries in question. Notwithstanding the President's authority to negotiate such agreements, it is difficult to justify them as a means of countering unfair trade or as a substitute for escape clause measures when many of the affected products never faced trade law remedies.

In this regard, the new policy showed an incompatibility of means and ends. If the goal was to protect against unfair trade, why not use the trade laws to determine violations and assess appropriate penalty duties? The ultimate irony of the new VER program was that, while touted as a means to protect against unfair trade practices, it did not allow the trade laws to work. The framework of the policy continued the tactics used in the 1982 US–EC pact, where trade laws only provided leverage to conclude broader quantitative trade restrictions. Furthermore, the reliance on market-sharing agreements to settle trade disputes contradicted the stated commitment to negotiate an end to trade-distorting practices (item 7), since guaranteed market shares, in satisfying producer interests, remove much of the incentive to achieve an open, market-driven trading system. The new agreements in effect indicated that any foreign government practices, including institutionalized subsidies, would be acceptable over the following 5 years as long as exporters honored their fixed market shares.

Another protectionist implication of the VER plan was that major exporters not already covered by VERs would have to be brought into the plan somehow, whether or not they were guilty of unfair trade practices. There was no evidence, for example, that Japan, South Korea and Canada were violating antidumping or countervailing duty laws. This is the apparent reason for threatening the use of section 301 (item 4 above) and for including diverted trade as a manifestation of an unfair trade practice (item 1), even though the United States had never imposed trade restrictions based on a section 301 finding of trade diversion (Coffield, 1982, p. 404). Presumably, if a foreign country limited its own imports of steel, thereby diverting steel exports to the United States, the President could take action under section 301 to impose trade sanctions against the foreign country guilty of trade-diverting protectionism. Reagan administration officials had suggested

that Japanese protectionism, for instance, had diverted steel exports to the United States, a charge the Japanese vehemently denied (*Chilton's Iron Age*, October 1984, p. 8). The USTR also threatened South Korea with a section 301 action, apparently because it was an alleged *source* of diverted trade.

It is clear that the role of section 301 in the steel VER program was to provide a convenient threat to pressure recalcitrant exporters into an agreement. The statute gives the President broad powers to unilaterally impose trade restrictions, based on USTR recommendations, for whatever he deems an 'unfair' foreign trade practice. In the cases of Japan and Korea, the charges seemed at best tenuous, and were in principle hypocritical, since United States protectionist policies have also diverted trade (as would the new VER plan). Yet these countries' large trade surpluses with the United States limited their diplomatic leverage in the matter, and the VER program became an attractive alternative.

The political economy of the VER agreement described earlier, whereby exporters face the threat of alternative protectionist action if they refuse and the benefit of an export price premium if they accept it, therefore ensured that most exporters would eventually submit to negotiated export limits. By late December 1984, even hold-outs such as Japan and Korea came around. Of the exporting countries targeted for VER negotiations, only Argentina refused to conclude an agreement. Table 7.2 summarizes the results of the VER negotiations, comparing previous import penetration levels by country with the new permissible market shares.

Despite the apparent success of the Reagan Administration's new steel trade policy in negotiating export restraint, serious difficulties, symptomatic of all such policies, remained. In so far as the VERs were indeed *not* global in their coverage of steel trade to the United States, the trade controls were still likely to 'leak,' and US trade officials soon revised their estimate of import penetration under the plan from 18.5 to 20.5 per cent. The Canadian steel industry, because of its close and complicated ties to US steel user markets, could not be forced to submit to a regulated and fixed VER limit. This left open the possibility of covert transshipment by other steel exporters through Canada, reports of which had already surfaced (*American Metal Market*, 21 December 1984, p. 8). Other sources of leakage included several small producers outside the VER framework, such as Sweden, Austria, Taiwan, several East European countries, Venezuela and other Latin American countries, many of which were planning to expand steel production and exports. In an attempt to plug these potential sources of import 'disruption' and perhaps encourage further negotiated export restraint, American steel producers filed unfair trade suits against

Table 7.2 *The 1984 steel pact agreements*

(a) *Market shares compared with pact agreement allocations*

| | Market Shares | | | Pact |
	1982	1983	1984[a]	Agreement
Japan	6.3	5.1	6.7	5.8
Korea	1.4	2.1	2.4	1.9
Brazil	0.8	1.5	1.4	0.8
Spain	0.7	0.7	1.6	0.67
South Africa	0.7	0.7	0.7	0.42
Mexico	0.1	0.8	0.9	0.36
Australia	0.2	0.2	0.3	0.18
Argentina	0.2	0.3	0.3	NA
Finland	0.2	0.2	0.4	NA
Canada	2.4	2.9	3.2	NA
EEC	7.3	4.9	5.9	5.9[b]
Total	21.8	20.5	26.1	—

Source: Commerce Department and US Trade Representative.
[a] First nine months
[b] 1982 US–EEC arrangement.

(b) *Semi-finished steel: pact import tonnage limit*

Country	Pact import tonnage limit (tons)
Brazil	700,000
Japan	100,000
Korea	50,000
Spain	50,000
South Africa	100,000
Australia	40,000
Mexico	100,000

Source: US Trade Representative.

several of these countries in late December 1984 (*American Metal Market*, 24 December 1984, p. 1, p. 24).

The pipe-and-tube controversy
In addition to a comprehensive set of trade restrictions, American steel producers also demanded the resolution of another unsettled issue: the non-binding 'gentleman's agreement' by EC steel producers in 1982 to limit pipe and tube shipments to 5.9 per cent of the US

market annually. This was the only major product category for which there was no enforcement mechanism in the 1982 US–EC steel pact. In the absence of binding restraint, EC pipe and tube shipments had reached 8.1 per cent of the US market in 1983 and had swelled to 14.4 per cent in the first eight months of 1984. The controversy over the EC's 'betrayal' of the agreement provides a useful insight into the nature of the bureaucratic machinery needed to make 'voluntary' export restraint work.

The root of the problem lay in the refusal of the major EC exporters of pipes and tubes in West Germany to submit to a structured export restraint program under the 1982 steel pact. Not only would a strict VER limit impose disproportionate German cutbacks in exports of these items as part of EC 'burden sharing' in the steel crisis, but the Commerce Department and ITC investigations in 1982 had actually dismissed most unfair trade cases against German steel producers. In their final determination no penalty duties on pipe and tube imports were recommended. The competitive German pipe and tube produc-ers therefore balked at the inclusion of their product in the pact, a situation that in fact threatened the entire agreement, as noted earlier. US producers, in the meantime, insisted on comprehensive controls. EC and US officials eventually papered over this dispute by agreeing to an 'informal' VER on pipes and tubes that would be exempt from mandatory export licensing, the vehicle of enforcement.

It is surprising that, in a continually soft steel market, anyone could have expected such an agreement to work. A VER agreement cannot work without a cartel-like arrangement that assigns specific export shares to each producer (see Jones, 1984, p. 89). In the absence of fixed market shares, each individual firms seeks to maximize profits by equating the marginal cost of production with marginal revenue at each market opportunity. Any effective plan of export restraint, in contrast, must force producers to accept a fixed share of the market, in which the firm's marginal revenue, equal to the collectively induced cartel price, *exceeds* marginal cost. Without strict enforcement, each cartel member therefore has an incentive to surpass its market share, leading, in the case of a VER, to expanded exports and lower prices. It was preposterous to expect a VER by so-called 'gentlemen's agree-ment' to restrain exports when the producers, naturally competing with each other for customers, were not forced to collude. The 'be-trayal' of the informal arrangement was thus the manifestation of competition among German and other EC producers for US custom-ers rather than the allegedly conscious and concerted effort by EC producers to disrupt the US market.

On 20 November 1984, the Reagan Administration none the less imposed a draconian measure, the embargo of all EC pipe and tube

shipments to the United States, in order to force EC officials to negotiate a binding VER arrangement. Trade legislation passed in October 1984 gave the President authority to unilaterally enforce the original 1982 figure of 5.9 per cent in the absence of a negotiated agreement. After several weeks of bitter wrangling among EC members over the allocation of internal market shares, the EC finally agreed to a 7.6 per cent market share for pipes and tubes on January 4, 1985. More than any other agreement of its type, the binding pipe and tube VER showed clearly that 'voluntary' export restraint represents not so much co-operation as it does coercion and the suppression of competition, and that it *forces* exporters to form a cartel.

Summary

As steel trade policy moved into more advanced stages of protectionism in the United States and the EC, the dynamic of the policymaking process made market-driven solutions to the industry's problems more difficult. By the early 1980s, government intervention to protect the industry from imports had severely distorted the incentive structure of adjustment, increasing the reaction threshold of steel firms that determines the speed of response to market conditions. Government policies and industry planning had failed to directly confront the vulnerability factors that lay at the root of market decline, leaving the industry in a state of continued crisis. This, in turn, led to a perpetual call for trade protection in both the United States and the EC.

Previous protectionist policies increased resonance among government policymakers to the industry's plight, increasing the steel industry's political influence and lowering the requisite threshold level of influence needed to bring about trade restrictions. In the United States, the failure of the TPM to curb steel imports once EC producers replaced the Japanese as the major source of market 'disruption' put trade officials in the position of designing new trade restrictions to plug the holes of the unsuccessful policy. When the resulting US−EC steel pact created yet more 'leaks' in the system, attracting imports from other, unrestrained countries, trade officials once again took the opportunity to try to correct a policy that had failed by constructing a more elaborate system of steel pacts. In the EC, the continued state of crisis and the sluggish implementation of government-directed restructuring caused the severe trade restrictions adopted in 1978 to become even more firmly entrenched, since imports posed a constant threat to the Community's *dirigiste* plans to manage the crisis. In this manner, the European Commission actually internalized the effects of trade

restrictions into their own policy goals, further dimming the prospects for a break in the protectionist cycle.

Other factors tended to enhance the resonance of policymakers to protectionist sentiment. The failure of earlier protectionist policies to return the industry to competitive health tended ironically to increase pressure for more protection. In the United States, the steel industry, sheltered from the full force of international competition for many years, was particularly vulnerable to the recession of the early 1980s and incurred huge financial losses during this period. Yet the industry's continued poor state of health set the stage for another successful plea for protection as policymakers perceived the danger of massive plant shutdowns. Steel producers had by this time refined their strategy of pursuing trade protection to a fine art, exploiting all possible political, legal and administrative avenues available to them, increasing their chances of success. They were aided during this period by a heightened protectionist mood in the United States, fuelled at first by the recession, but later by increasing trade deficits and the rise of the US dollar on currency markets. In the EC, the continued depression in the steel industry also contributed to protectionist sentiment, but this was offset to a certain degree by growing dissatisfaction with failed government subsidy programs for national steel industries. However, the course of steel trade policy at this point remained in the firm grip of the supranational European Commission, which showed no signs of abandoning its highly restrictive trade controls.

In general, steel trade policy developments from 1978 to the mid-1980s revealed the tendency of protectionism to feed on itself, enclosing policymakers in a spiral of increasing government intervention in the industry. A large portion of the unfinished business of adjustment to the fundamental change in competitive factors in the world steel industry therefore still lay ahead for producers in these countries.

Notes

[1]   See OJEC, no. L291, 1980, p. 1, art 7(2). The ratio is determined by the pattern of the firm's steel deliveries to its customers in the EC and in third (non-EC) countries that was established from 1977 to 1980.

[2]   Assume, for example, a production quota of 100 units, a delivery ratio of 0.9 and foreign exports of 5 units. The maximum allowable amount of EC internal deliveries, x units, will be determined by $0.9 = x/(x+5)$. The solution $x = 45$ shows that, even though the production quota is set at 100 units, the firm may not deliver more than 50 units in all. By increasing exports to 10 units, the new solution for allowable EC internal deliveries is $x = 90$, permitting the firm to fill its entire production quota.

[3]   See *Wall Street Journal*, 22 October 1982, p. 3. Off-the-record comments by a former American government official indicate that the agreement was actually negotiated in

secret between US and Japanese officials during the latter stages of the Tokyo Round of multilateral trade negotiations.

4 See the Trade Act of 1974, section 203, which gives the President the authority to negotiate orderly marketing agreements. None the less, in the latter stages of the negotiations, both the European Commission and the Commerce Department consulted the Justice Department Anti-Trust Division to confirm the legality of the agreement.

5 *The Economist*, 23 October 1982, p. 50. The principles of 'collective good' and 'solidarity' were invoked by the European Court of Justice in its conviction of steel firms from the Brescia area in Italy for illegal price-cutting. See Trainer, 1980, pp. 559–611.

6 The ITC's final determination of injury, which would have made the duties official, was therefore never published, despite the fact that the Commissioners had reached an official decision. Commissioner Stern, resenting the summary termination of the case without promulgation of the ITC's findings, issued her own views on the case in an annex to a separate determination. See ITC, 1982b.

7 The ITC recommended relief for (1) semi-finished steel, (2) plates, (3) sheets and strip, (4) wire and wire products, and (5) structural shapes and units. It rejected relief for wire rods, railway products, bars and pipes and tubes. See ITC, 1984c, pp. 9–23.

8

Conclusions and Outlook: The Implications of Non-Adjustment

Resistance to adjustment

The drastic changes in the world steel market in the postwar period have challenged the established steel industries of the United States and the EC to adjust by modernizing, cutting costs, specializing in a more limited range of steel product lines, contracting and closing down marginal, uncompetitive steel capacity. Yet resistance to the often painful process of adjustment has been strong. Protection of national industries from import competition has strong and deep historical roots, and the mystique of steel as a symbol of international economic strength, to be protected from the onslaught of foreign producers, remains to this day undiminished. It is in this atmosphere of crisis that import competition has become the focus of policies designed to 'save' the steel industry. Market pressures, exacerbated by severe international competition, have indeed caused much pain in the industry and disruption in steel-producing regions. Yet the need for adjustment remains; it is a part of the process of economic change to which the industry and the world economy as a whole must adapt in order to provide the conditions for further economic growth.

The sources of change and competitive shifts in the world steel industry have in fact many roots; trade is not the sole — or even the primary — reason for the adjustment crisis. As was shown in Chapter 4, the secular decline in steel demand and a switch towards substitute materials in advanced industrialized countries have been a major source of decline. Domestic mini-mills have begun to supplant the large integrated producers in many product lines. Technological innovations and their spread to less developed countries, lower shipping costs and shifting input cost factors have changed the pattern of steelmaking competitiveness. The decidedly uncompetitive cost structure of some steel firms, in conjunction with poor managerial performance and

damaging government intervention, have forfeited even more competitive advantage among established producers. In view of the myriad of causes of competitive change in the world steel industry, increased imports in the United States and the EC have represented a symptom, rather than a cause, of the underlying adjustment problem.

For all the arguments marshalled forth by the steel industry and by governments for the need to control 'disruptive' imports, however, a compelling case for steel trade restrictions has failed to emerge. Indeed, warnings of the dire consequences of increased imports have found little support in the economic facts of international steel trade. The American industry has asserted, for example, that increased import penetration and declining domestic steel production would expose American steel consumers to monopoly prices during steel shortages, citing the 1973–4 boom period as evidence. Yet a comparison of the present values of the cost of importing steel at premium prices during a shortage and at low prices during periods of oversupply on world markets shows that trade has *not* raised the average price of steel over the business cycle. In an increasingly competitive world market for steel, marked by the wide dispersion of steel supply, the argument that trade restrictions somehow protect consumers acquires the hue of fantasy.

The claims that 'unfair' trade have severely distorted the pattern of steel competitiveness and exports also have a hollow ring. While it is true that a uniform, consistent and *non-protectionist* application of unfair trade laws is a necessary part of a politically viable international trading system (a point to be discussed below), allegations of unfair trade without the requisite evidence cannot act as a legitimate basis for protectionist polices *outside* the relevant trade laws themselves. It is evident, for example, that governments have played an important role in directing the development of national steel industries in the postwar period, and have become increasingly involved in the ownership of steel capacity in some countries, but there is no convincing evidence that the governments of major steel exporting countries have substantially enhanced their national industries' basic competitive position. In Europe, in fact, there is evidence that government intervention has actually harmed steel firms there. The available evidence suggests that the *net* impact of government programs, through the price of exported steel, on the steel industries of importing countries is marginal. In those cases where subsidization or dumping has been proven, trade law remedies provide the appropriate means of redressing the violation. If the trade laws themselves were allowed to operate without interference, the final impact on steel trade would be considerably smaller than that of the protectionist export restraint agreements that have often replaced their enforcement. In view of the declining competitive

position of integrated producers in the United States, who have been most aggressive in such allegations of 'unfair' trade, this issue is in economic terms largely irrelevant. It is the set of economic factors enumerated above, not foreign subsidies, that has caused the decline of the steel industry in the United States.

At the same time, warnings of the imminent extinction of the steel industries of the United States and the EC have been exaggerated. In general, these countries are, contrary to protectionist claims, not 'de-industrializing' (see R. Lawrence, 1983). The viability of national steel industries in these major industrial areas appears to be secure, despite the turmoil of adjustment, and steelmaking capacity and supply sources for strategic and national security purposes are not in jeopardy. However, in the absence of a dramatic surge in world steel demand – now regarded as an unlikely event – it is clear that market pressures will cause some further plant closings (Crandall, 1981, p. 142). Carbon steel output in certain product lines is likely to shift increasingly towards the mini-mill sector, with these national steel industries as a whole concentrating more on specialty steels and high value-added product lines. The focus of basic steelmaking is likely to shift increasingly towards the newly industrializing countries (Walter, 1979, p. 182).

The continued decline of the steel industry, in spite of all the programs of protection since 1968, has in fact caused steel firms to begin programs of restructuring. From the depths of the industry's depression, there have been significant signs of increased efficiency and productivity, quality improvements and cost reductions (Mueller, 1984, pp. 111–12). Hogan (1984, p. 138–40) predicts that the obsolete plant and equipment that have burdened the US steel industry will eventually be scrapped, perhaps by the late 1980s, while noting that steelmaking capacity in the United States and the EC has declined since 1980. If this process continues, the smaller size of the steel industry will ease the burden of adjustment, as available funds for modernization can be focused on facilities that can be competitive on world markets in future years. Even in Japan, the world efficiency leader among major steel producing nations, a process of restructuring has taken place, as steelmaking capacity has dropped from 150 million tons in 1980 to an estimated 120 million tons in 1985.

In this regard, it has become increasingly evident that protectionism has failed to halt the decline of the steel industries it was supposed to shelter; it has merely delayed an inevitable process of adjustment. The promising signs of gradual improvement in the highly protected American and EC markets provide an even more compelling reason to dismantle steel trade restrictions: No further incentives to delay or avoid adjustment should be placed in their path. Protection in the form

of subsidies appears to be easiest to remove. Public tolerance of expensive government bail-outs of the steel industry, especially in the Community, has declined considerably, and some countries, such as the United Kingdom and France, have begun to eliminate subsidy programs that have been in place for many years. The resulting dissatisfaction among displaced steelworkers, accompanied by violence in some cases, shows that the costly policies of protection and public support have not even been able to achieve the basic goal of social peace. The protected steel industries therefore appear, despite the distortions of protectionism, to be moving slowly towards an economic solution. It is in the economic and political interest of the countries involved to resist further measures that would only draw out the painful period of adjustment even more.

The cost of protection and the steel trade policy crisis

Despite the evidence that the market system in steel trade poses no threat to the welfare of advanced industrial countries, policies of protectionism have prevailed since the late 1960s. Such policies have allowed firms to delay the process of market-driven adjustment by extracting from consumers an implicit subsidy in the form of price increases, which the trade restrictions have transferred to both domestic and foreign steel producers. In the debate over steel trade policy, this transfer has remained at the heart of every protectionist campaign, notwithstanding the claims of 'unfair' trade, national defense needs, or the danger of de-industrialization. A study of the impact of steel trade restrictions on steel prices in the United States estimates that the VRA agreements increased import prices by 6.3 to 8.3 per cent, pushing up overall domestic steel prices by 1.2 to 3.5 per cent. The TPM, according to the study, increased import prices by 10.3 per cent and average domestic steel prices by 2.7 per cent. Although the percentages themselves do not appear to be large, the massive volume of steel shipments in the United States suggests that the trade restrictions have transferred billions of dollars from consumers to producers, the TPM alone transferring $3 billion (Crandall, 1981, pp. 105, 111, 137).

Yet the direct consumer and efficiency cost of trade restrictions in steel only begin to describe the detrimental impact that steel protectionism has had on national economies and the world trading system. Much of the damage has occurred because of the peculiar course of steel trade policy since the adjustment crisis began. The long-standing crisis in steel trade policy has been identified in this study as a result of three interacting elements corresponding to the cause, generation and trade

effects of the particular protectionist policies used. First, shifts in competitiveness on international steel markets have given rise to protectionist sentiment as import-competing firms in the United States and the EC have resisted adjustment. The painful cuts in capacity needed to return the industry to economic health have not been forthcoming, as the high political profile of steel has allowed the industry to achieve its protectionist goals.

The second element of the crisis has been formed by the policy choices of governments in accommodating protectionist sentiment. In order to avoid violations of trade agreements and contain possible retaliation, trade authorities in the United States and the EC have turned to systems of induced export restraint, including VERs, threats of trade litigation and trigger pricing schemes. Quantitative 'voluntary' restraint agreements have emerged as the most popular device of this type of trade restriction, since they allow exporting suppliers to raise export prices, import-competing firms to raise domestic prices and governments to discriminate against individual 'disruptive' suppliers, thereby limiting the scope of possible retaliation. Consumers in the importing country are the principal losers in the arrangement, but are generally not represented in the negotiations, which are usually held out of public view and are dominated by producer interests. In keeping with the letter (if not the spirit) of GATT restrictions on quantitative and discriminatory *import* barriers, such systems of 'voluntary' restraint have provided a convenient channel of protectionist policymaking through which industry demands for trade restrictions can be accommodated.

Finally, the policies of export restraint themselves have had the effect of diverting exports to other available markets, refocusing the import problem on the 'rebound' market and proliferation protectionist policymaking worldwide. The resulting suspicion and instability in trade relations are in fact the basic reasons that the GATT in principle opposes such discriminatory devices through its most-favored-nation clause. In the case of steel, American and EC trade policies have been following a pattern of spiralling, 'rebound' protectionism ever since the United States introduced VRAs in 1968. Table 8.1 traces the protectionist cycle in steel trade relations since this time. Through a combination of trade diversion and policy imitation, a concatenation of discriminatory steel trade policies has resulted. The first VRAs imposed by the United States were followed by the first VERs imposed by the EC on Japan; the ensuing trigger price mechanism (TPM) of the United States then led directly to the introduction of the basic price system of the Community. Leakage problems caused the TPM to yield to the steel export restraint pact between the United States and the Community, but further leakage resulted in renewed protectionist

Table 8.1 *Export Restraint and protectionist cycles in the steel trade*

Type	Importing area	Main target countries	Years of coverage	Rebound protectionism	Main source of leakage (unrestrained exports)	Attempts to close leakages
VER (a.k.a. VRAs)	USA	EEC, Japan	1968–74	Japanese export diversion to EEC→ EEC concludes first VER with Japan (1971)	'Renegade' EEC/ Japanese firms, UK, some LDCs	Tighten coverage, include UK in second VRA (1972–4)
VER	EEC	Japan	1976	Japanese export diversion to US → US TPM	Several non-restraining countries	VER/BPS, introduced 1978
TPM, informal VER	USA	Japan	1978–82 (suspended Mar.–Oct. 1980)	EEC fear of export diversion EEC imposes VER/BPS	EEC	Export arrangement W/EEC, October 1982
VERs–BPS	EEC	Most supplying countries	1978–present	Some LDC and other exports diverted to US→ protectionist pressure in US	Some LDCs, other countries	Increasing VER coverage of all suppliers
Export arrangement	USA	EEC	1982–5	EEC exports diverted to home markets → EEC tightens VERs	LDCs, other non-restraining countries	Protectionist pressure for global import quota
Planned global VERs	USA	NICs Japan	1984–9	Tighter EC VERs?	Non-restraining countries?	Global steel export Cartel?

pressure and the conclusion of a broad set of new export pacts with nearly all suppliers to the United States. In this manner, the cycle of protectionism in steel trade has generated a self-sustaining policy crisis.

The cartelization of the EC's steel industry has in this regard created a special set of adjustment problems. While the stated purpose of the crisis measures has been to restructure the industry through planned capacity cuts and modernization, experience suggests that the sought-after adjustment can only occur in a timely fashion if all firms are finally denied the opportunity to, for example (1) hold out for higher production quotas; (2) press for higher minimum prices; or (3) expect continued hermetic protection from imports. Furthermore, the elimination of competition ostensibly needed to pursue 'orderly' restructuring policies has applied not only to imports. Since the Community's steel industry would certainly otherwise be highly competitive internally, the cartel policies of the European Commission make it necessary to suppress market penetration by efficient Community firms as well. These policies have allowed over-capacity to persist, which has caused EC officials to devise a system of quotas and internal delivery restrictions that encourages exports. This, in turn, generated increased protectionist sentiment in the United States.

The most serious implication of official intervention in the EC steel market for international trade policy, however, is that of subsidization. Heavy government involvement in the steel industries of several EC member countries has not only delayed adjustment, it has also tainted trade relations. A key element in the achievement of trade liberalization under the GATT has been the acceptance among its members of basic rules of conduct in commercial policy, including provisions restricting the use of government subsidies that distort international trade patterns. Although there is little evidence that subsidies themselves have caused significant distortions to steel trade or injury to the established steel industries of the United States and the EC, the use of massive state aids to achieve domestic employment and regional policies has cast a pall over steel trade relations. The national character of steel industries, combined with the ever-present suspicions of foreign predatory motives, has made the subsidy issue the major stumbling block to improved steel trade relations. In the case of the US—EC steel agreement, the accusations of subsidization among EC producers have had the added perverse effect of harming the efficient, non-subsidized firms, which according to the principle of burden-sharing have had to reduce exports along with the subsidized firms. It is safe to say that both the EC's internal steel crisis and the world steel trade policy crisis will persist as long as massive subsidization within EC countries continues.

A framework for resolving the policy crisis

Is, however, a comprehensive solution to the crisis possible? The political will among governments to pursue an economic solution based on market adjustment is certainly not in evidence, based on policy experience since 1968. The course of policies so far has in fact moved in the opposite direction: towards a tighter system of 'voluntary' restraint arrangements covering larger and larger quantities of world steel trade. This trend has its roots both in the continuing adjustment crisis accompanying protection and in the tendency of discriminatory trade restrictions to generate a spiralling cycle of protectionist responses worldwide. In addition, non-universal export restraint cannot succeed longer than the time it takes for non-participating or new suppliers to cash in on the premium export price thereby established for the market of destination. As the steel trade policy crisis has worsened, consideration of a universal system of orderly marketing agreements has therefore begun to appear. The conclusion of export restraint agreements by the United States with most of its steel supplying countries, in combination with the Community's comprehensive network of VERs, has already established the framework for such an arrangement. A worldwide steel export cartel would be the logical culmination of protectionism via export restraint.

Thus the protectionist cycle in steel trade relations has peaked once again, with movement towards a resurrection of the International Steel Cartel of the 1920s and 1930s, this time under direct government control. As the historical review of Chapter 2 showed, the frustration of steel producers with the unpleasant effects of increasing international competition provides ample motive for the conclusion of collusive, market-sharing agreements. The mistake made by many governments during the time of the earlier cartels was to regard, as if by wishful thinking, such arrangements as a sign of international co-operation, where in reality they were nothing more than price-fixing schemes designed to accommodate producer interests. The lesson apparently learned by the signatories of the GATT in the early postwar period but forgotten later was that restrictions on international trade, especially through the elimination of competition, are detrimental not only to economic welfare and growth, but also to an orderly and peaceful *system* of international trade relations. The failure of international steel cartels to establish a viable 'order' in steel trade in the prewar period − and its probable link to the deterioration of trade and political relations prior to the war itself − should have given policymakers an indication of the consequences of a renewed cartel. Yet the EC established a cartel during the adjustment crisis period of the 1970s, a failure in its own right, and trade policy measures by the

United States and the Community have moved inexorably towards similar arrangements covering export supply.

The implications of further cartelization for worldwide economic growth and welfare would be potentially disastrous, however, not only in terms of misallocated resources and loss of consumer welfare in steel consumption, but also in terms of the long-term effects of the resulting trade distortions. With international competition suppressed on world markets, economic growth in both industrialized and industrializing countries would decline, productivity and technological advancement in the steel industry would be hindered, and steel consumers would face permanently higher prices. Those countries with an otherwise increasing competitive advantage in steel production would be particularly damaged by such a rigid market-sharing arrangement. The suppressed growth of these countries would feed back to the detriment of the economies of industrialized countries, whose export markets would suffer as a conseqence.

The advanced state of government intervention, however, both in terms of protectionism, subsidization and other government involvement, appears to rule out a *laissez-faire* solution to the problems of world steel trade. It is clear that some form of negotiations are necessary to resolve the crisis, but the area of greatest promise lies not in carving out 'fair' market shares, but in re-asserting the principle of non-discrimination and the integrity of national trade laws and in allowing competitive advantage to determine the structure of trade. Perhaps the most damaging result of the steel trade policy crisis so far has been its role in subjecting trade laws to protectionist manipulation. A new consensus is needed among trading nations, particularly the United States and EC members, on general rules and standards regarding government intervention in the steel industry, in steel trade and in the use of trade laws. Special emphasis should be placed on rules regarding production subsidies, as well as on specific timetables to end subsidies that distort steel trade patterns. Once such standards are established, it may be possible to allow a uniform and consistent application of antidumping and unfair trade practice laws and article XIX of the GATT (the escape clause) to perform their intended function as safety valves for protectionist pressure.

Ultimately, it will be necessary as well to commit trading nations to eliminate VER and other market-sharing arrangements as a means of settling trade disputes. A possible first step in this direction would be the removal of quantitative restrictions and their replacement with non-discriminatory global tariffs. This would then provide the basis for multilateral negotiations to systematically reduce tariff barriers over time. The consumer, as well as the stability of trade relations in general, would benefit from the return to a multilateral liberalization

of steel trade and the uniform enforcement of national trade laws that are consistent with the GATT. A pattern of world steel trade based on natural market advantages may then be possible.

The elements of a workable and lasting solution to the steel trade policy crisis point, in the final analysis, to the need for reform in world trade relations in general. While it may appear at first that the trading system faces a multitude of problems, from steel, textiles and automobiles to agriculture, electronics and high-technology goods, from north—south relations to the need for a 'level playing field,' they are in large part the manifestations of two more basic problems: discrimination and the lack of transparency in trade policy. The value of a close examination of the course of steel trade restrictions, with its record of induced export restraint and spiraling protectionism, thus lies in its ability to document the futility of international market-sharing measures as a means of solving the basic need for adjustment to changing economic conditions. In so far as instruments of induced export restraint have succeeded in reducing import competition, they have delayed adjustment, discriminated against low-cost suppliers and spread protectionism by diverting exports towards other markets. In so far as they have failed to reduce import competition, they have caused policymakers to adopt even more restrictive and market-distorting forms of protectionism, sparking new international trade policy crises. In general, such measures have fostered a system of perpetual laxity in competitiveness as well as a constant state of suspicion and crisis among trading partners. Yet their political attractiveness as a negotiated means of accommodating the interests of established producers — to the systematic exclusion of consumer interests — provides a continuing incentive for industries and governments to pursue them. Once established as an accepted and acceptable instrument of trade policy, their use could easily spread further, compounding their damage to the world economy.

Steel protectionism therefore represents, indeed, a special case in international trade policy. The nearly mythical significance that governments have attached to the goal of a strong national steel industry, providing independence from an uncertain and hostile world, has given the industry tremendous political influence and the ability to achieve protection from foreign competition. Yet the peculiar story of the steel trade policy crisis presents perhaps a more important lesson to governments and a concerned public, for the far-reaching schemes of steel export restraint and cartelization threaten to undermine the entire system of trade rules so painstakingly achieved in the wake of the terrible destructiveness of economic depression and world war, events directly linked with the chaotic pattern of aggressive protectionist policies of the 1930s. The ability of the new protectionist measures

to stunt economic growth and destabilize trade relations in this same manner provides a compelling case to seek a renewal of a system of trade based on the pattern of market-driven international competitiveness. For the steel industry and for the world economy as a whole, this is the only truly viable solution.

Bibliography

Adams, W. and Dirlam, J. B. (1966), 'Big steel, invention and innovation,' *Quarterly Journal of Economics*, May, pp. 167–89.

Adams, W. and Dirlam, J. B. (1977), 'Import competition and the Trade Act of 1974: a case study of section 201 and its interpretation by the International Trade Commission,' *Indiana Law Journal*, vol. 52(3), pp. 535–99.

Adams, W. and Mueller, H. (1982), 'The steel industry,' in W. Adams (ed.), *The Structure of American Industry* (New York: Macmillan), pp. 73–133.

AISI (American Iron and Steel Institute) (annual), *Annual Statistical Report* (Washington, DC: AISI).

AISI (1984), 'Steel Imports News' (press release), 18 September (Washington, DC: AISI).

Anderson, R. G. and Kreinin, M. E. (1981), 'Labor costs in the American steel and auto industries,' *The World Economy*, vol. 4(2), pp. 198–208.

Arrangement Concerning Trade in Certain Steel Products Between the ECSC and the United States, 21 October 1982.

Ault, D. (1973), 'The continued deterioration of the competitive ability of the U.S. steel industry: the development of continuous casting,' *Western Economic Journal*, March, pp. 89–97.

Barcelo, J. J. (1979), 'The antidumping law: repeal it or revise it,' in S. M. Harris (ed.), *Antidumping Law: Policy and Implementation, Michigan Yearbook of International Legal Studies*, Vol. I (Ann Arbor: University of Michigan Press), pp. 53–96.

Barnett, D. F. and Schorsch, L. (1983), *Steel: Upheaval in Basic Industry* (Cambridge, Mass.: Ballinger Publishing Company).

Baumann, H. G. (1974), 'The relative competitiveness of the Canadian and US steel industries, 1955–70,' *Economia Internazionale*, vol. 27 (Feb), pp. 141–56.

Berg, R. C. (1982), 'Petitioning and responding under the escape clause: one practitioner's view on how to do it,' *North Carolina Journal of International Law and Commercial Regulation*, vol. 6, pp. 407–26.

Berglund, A. and Wright, P. G. (1929), *The Tariff on Iron and Steel* (Washington DC: Brooking Institution).

Blackhurst, R., Marian, N. and Tumlir, J. (1977), *Trade Liberalization, Protectionism and Interdependence* (Geneva: General Agreement on Tariffs and Trade).

Bollino, C. A. (1983), 'Industrial policy in Italy: a survey,' in F. G. Adams and L. R. Klein (eds.), *Industrial Policies for Growth and Competitiveness* (Lexington, Mass.: Lexington Books).

Brown, F. J. (1977), 'Letter from Local 1013, United Steelworkers of America,

to Congressman J. Buchanan, September 13,' in *U.S. Congress, Committee on Ways and Means, World Steel Trade: Current Trends and Structural Problems*, Serial 95–37 (Washington, DC: US Government Printing Office).

Burn, D. L. (1961), *The Economic History of Steelmaking, 1867–1939* (Cambridge: Cambridge University Press).

Carr, J. C. and Taplin, W. (1962), *History of the British Steel Industry* (Cambridge, Mass.: Harvard University Press).

Coffield, S. (1982), 'Using section 301 of the Trade Act of 1974 as a response to foreign government trade actions: when, why and how,' *North Carolina Journal of International Law and Commercial Regulation*, vol. 6, pp. 381–405.

Comptroller-General of the United States (1974), *Economic and Foreign Policy Effects of Voluntary Restraint Agreements on Textiles and Steel* (Washington, DC: US Government Printing Office).

Comptroller-General of the United States (1981), *New Strategy Required for Aiding Distressed Steel Industry* (Washington, DC: US Government Printing Office).

Congressional Budget Office (1984), *The Effects of Import Quotas on the Steel Industry* (Washington, DC: US Congress).

Corden, W. M. (1974), *Trade Policy and Economic Welfare* (Oxford: Oxford University Press).

Council on Wage and Price Stability (1977), *Report to the President on Prices and Costs in the United States Steel Industry* (Washington, DC: US Government Printing Office).

Crandall, R. W., (1981), *The U.S. Steel Industry in Recurrent Crisis* (Washington, DC: Brookings Institution).

Dale, R. (1980), *Anti-dumping Law in a Liberal Trade Order* (London: Macmillan).

Davignon, E. (1979), 'Restructuring in Europe,' in *Steel Industry in the Eighties*, Proceedings of the International Conference of the Metals Society (London, September 1979).

Department of Commerce (1984), 'Foreign import restraints and unfair practices in steel' memorandum September 18, (Washington, DC: US Department of Commerce).

De Witt, F. (1983), 'French industrial policy from 1945–1981: an assessment,' in F. G. Adams and L. R. Klein (eds.), *Industrial Policies for Growth and Competitiveness* (Lexington, Mass.: Lexington Books).

Farrer, T. H. (1886), *Free Trade versus Fair Trade* (London: Cassell).

Federal Trade Commission (1977), *Staff Report on the United States Steel Industry and Its Rivals: Trends and Factors Determining International Competitiveness*. Staff Report of the Bureau of Economics (Washington, DC: US Government Printing Office).

Finger, J. M., Hall, H. K. and Nelson, D. R. (1982), 'The political economy of administrative protection,' *American Economic Review*, June, pp. 452–66.

Frank, E. (1984), Speech delivered at the Eastern Economic Association Conference, New York, 12 March 1984.

Goldstein, G. (1912), 'Der deutsche Eisenzoll, ein Erziehungszoll,' *Volkswirtschaftliche Zeitfragen*, Jahrgang 34, Heft 4 (268).

Gothein, G. (1904), 'La Réglementation Internationale des Droits de Douane sur les Fers,' *Revue Économique Internationale*, vol. 1, pp. 483–527.

Gray, H. P. (1973), 'Senile industry protection: a proposal,' *Southern Economic Journal*, vol. 39, pp. 569–74.

Grubel, H. (1982), *International Economics*, 7th edition (Homewood, Ill.: Richard O. Irwin).

Hamilton, A. (1971), *Report on Manufacturers*, reprinted in F. W. Taussig (ed.), *State Papers and Speeches on the Tariff* (1968) (New York: Burt Franklin).

Hexner, E. (1943), *The International Steel Cartel* (Chapel Hill: University of North Carolina Press).

Hogan, W. T. (1971), *Economic History of the Iron and Steel Industry in the United States*, 5 Vols. (Lexington, Mass.: Lexington Books).

Hogan, W. T. (1983), *World Steel in the 1980s: A Case of Survival* (Lexington, Mass.: Lexington Books).

Hogan, W. T. (1984), *Steel in the United States: Restructuring to Compete* (Lexington, Mass.: Lexington Books).

International Iron and Steel Institute (IISI) (1974), *Steel Intensity and GNP Structure* (Brussels: IISI).

International Iron and Steel Institute (1981), *World Steel in Figures* (Brussels: IISI)

ITC (1978), *Carbon Steel Plate from Japan*, AA 1921–179 (Washington, DC: US ITC).

ITC (1979), *Carbon Steel Plate from Taiwan*, AA 1921–197 (Washington, DC: US ITC).

ITC (1982a), *Certain Steel Products from Belgium, Brazil, France, Italy, Luxembourg, the Netherlands, Romania, the United Kingdom, and West Germany*, publication 1221 (Washington, DC: US ITC).

ITC (1982b), *Carbon Steel Wire Rod from Brazil and Trinidad and Tobago*, publication 1316 (Washington, DC: US ITC).

ITC (1984a), *Foreign Industrial Targeting and its Effects on US Industries Phase II: The European Community and Member States*, publication 1517 (Washington, DC: US ITC).

ITC (1984b), *Operation of Trade Agreement Program*, publication 1535 (Washington, DC: US ITC).

ITC (1984c), *Carbon and Certain Alloy Steel Products*, publication 1553 (Washington, DC: US ITC).

Johnson, C. (1982), *MITI and the Japanese Miracle: The Growth of Industrial Policy, 1925–1975* (Stanford: Stanford University Press).

Joliet, R. (1981), 'Cartelization, dirigism and crisis in the European Community,' *The World Economy*, vol. 3(4), pp. 403–45.

Jones, K. (1979), 'Forgetfulness of things past: Europe and the steel cartel,' *The World Economy*, vol. 2(2), pp. 139–54.

Jones, K. (1981), The political economy of voluntary export restraint and the incidence of trade diversion in steel import markets (Geneva: unpublished dissertation, University of Geneva).

Jones, K. (1984), 'The political economy of voluntary export restraint agreements,' *Kyklos*, vol. 37(1), pp. 82–101.

Kawahito, K. (1981), 'Japanese steel in the American market: conflict and causes,' *The World Economy*, vol. 4(3), pp. 229–50.

Keeling, B. (1982), *The World Steel Industry: Structure and Prospects in the 1980s* (London: Economist Intelligence Unit Special Report, No. 128).

Kestner, F. (1902), *Die deutschen Eisenzolle, 1879–1900* (Leipzig: Duncker and Humbolt).

Kiers, L. (1980), *The American Steel Industry: Problems, Challenges, Perspectives* (Boulder, Col.: Westview Press).

Kiersch, G. (1954), *Das Internationale Eisen- und Stahlkartell* (Essen: Rheinisch-Westfalisches Institut für Wirtschaftsforschung).

Kreinin, M. (1985), 'Wage competitiveness in the US auto and steel industries,' in J. Adams (ed.), *The Contemporary International Economy: A Reader* (New York: St. Martin's Press), pp. 174–88.

Lawrence, P. R. (1983), *Renewing American Industry* (New York: Free Press).

Lawrence, R. Z. (1983), 'Is trade deindustrializing America?' *Brookings Papers on Economics Activity*, vol. 1, pp. 129–71.

Lehnardt, D. G. (1969), 'Executive authority and antitrust considerations in "voluntary" limits on steel imports,' *University of Pennsylvania Law Review*, vol. 118, pp. 105–28.

List, F. (1841), *Das nationale System der politischen Ökonomie* (Stuttgart: Cotta).

Lister, L. (1960), *Europe's Coal and Steel Community* (New York: Twentieth Century Fund).

Lloyd, P. (1977), 'Antidumping actions and the GATT system,' *Thames Essay No. 9* (London: Trade Policy Research Center).

Lowenfeld, A. (1979), 'Public controls on international trade,' *International Economic Law Series*, (VI) (New York: Matthew Bender).

McConnell, G. (1963), *Steel and the Presidency, 1962* (New York: W. W. Norton).

McConnell, J. W. (1943), *The Basic Teachings of the Great Economists* (New York: The New Home Library).

Mueller, H. (1982), 'A comparative analysis of steel industries in industrialized and newly industrializing countries,' *Conference Paper No. 72*, Middle Tennessee State University Conference.

Mueller, H. (1984), 'Protection and competition in the U.S. steel market: a study of managerial decision making in transition,' *Monograph Series, No. 30* (Tenn.: Middle Tennessee State University).

Mueller, H. and van der Ven, H. (1982), 'Perils in the Brussels–Washington steel pact of 1982,' *The World Economy*, vol. 5(3), pp. 259–78.

Office of Technology Assessment (1980), *Technology and Steel Industry Competitiveness* (Washington, DC: US Government Printing Office).

Organization for Economic Cooperation and Development (OECD) (1977), *The Iron and Steel Industry in 1975* (Paris: OECD).

Ovenden, K. (1978), *The Politics of Steel* (London: Macmillan).

Patenode, T. J. (1980), 'The new anti-dumping procedures of the Trade Agreements Act of 1979: does it create a new non-tariff trade barrier?' *Northwestern Journal of International Law and Business*, vol. 2, pp. 200–23.

Patrick, H. and Sato, H. (1982), 'The political economy of United States–Japan trade in steel,' in K. Yamamura (ed.), *Policy and Trade Issues of the*

Japanese Economy: American and Japanese Perspectives (Seattle and London: University of Washington Press).

Preeg, E. H. (1970), *Traders and Diplomats* (Washington, DC: Brookings Institution).

Putnam, B., Hayes, B. and Bartlett (1979), *The Economic Implications of Foreign Steel Pricing Practices in the US Market* (Washington, DC: American Iron and Steel Institute).

Reagan, R. (1984), 'Steel import relief determination,' (Memorandum for the U.S. Trade Representative), 18 September.

Report to the Commissioner Appointed under Provisions of Act 5&6 Vict. c.99 to Inquire into Operation of Acts and into State of the Population as to Education of Schools in the Mining Districts (1854) *House of Commons Command Paper, vol. XIX* (London: House of Commons).

Rodriguez, F. A. (1979), 'Gilmore: an antidumping proceeding as cost-price comparison,' in S. M. Harris (ed.), *Antidumping Law: Policy and Implementation, Michigan Yearbook of International Legal Studies*, Vol. I (Ann Arbor: University of Michigan Press).

Sakoh, K. (1983), 'Industrial policy: the super myth of Japan's super success,' *Asian Studies Center Backgrounder*, No. 3 (Washington, DC: Heritage Foundation).

Sandmo, A. (1971), 'On the theory of the competitive firm under price uncertainty,' *American Economic Review*, vol. 61, pp. 65–73.

Stanwood, E. (1904), *American Tariff Controversies in the Nineteenth Century* (New York: Houghton, Mifflin and Co.).

Stegemann, K. (1977), 'Price competition and output adjustment in the European steel market,' in H. Giersch (ed.) *Kieler Studies* No. 147 (Tubingen: J. C. B. Mohr Paul Siebeck).

Stocking, G. W. (1950), 'Testimony before Congress,' in *Subcommittee on Study of Monopoly Power*, 81st Congress, 2nd Session (Washington, DC: US Government Printing Office).

Stolper, W. and Samuelson, P. A. (1941), 'Protection and real wages,' *Review of Economic Studies*, November, pp. 58–73.

Taussig, F. W. (1915), *Some Aspects of the Tariff Question* (Cambridge, Mass.: Harvard University Press).

Taussig, F. W. (1931), *The Tariff History of the United States*, 8th edition (New York: Augustus M. Kelley).

Trainer, R. (1980), 'The concrete reinforcement bars case and the Davignon plan: judicial endorsement of the ECSC's crisis policies,' *The Journal of International Law and Economics*, vol. 14(3), pp. 559–611.

United Nations Economic Commission for Europe (annual), *Statistics of World Trade in Steel* (Geneva: United Nations).

US Congress (1968), *Steel Imports*, Staff Study of Senate Committee on Finance, Senate Document no. 107, 90th Congress (Washington, DC: US Government Printing Office).

US Tariff Commission (1928), *Iron and steel: a survey of the iron and steel industries and international trade of the principal producing and trading countries with particular reference to factors essential to tariff considerations, report no. 128*, second series (Washington, DC: US Government Printing Office).

United States Trade Representative (1984), 'Brock announces President's steel decision' (Press release), 18 September 1984.

Van Bael, J. (1978), 'The EEC antidumping rules – a practical approach,' *International Lawyer*, vol. 12, pp. 523–45.

Viner, J. (1923), *Dumping: A Problem in International Trade* (New York: Augustus M. Kelley).

Walter, I. (1979), 'Protection of industries in trouble – the case of iron and steel,' *The World Economy*, vol. 2(2), pp. 155–87.

Walter, I. and Jones, K. (1981), 'The battle over protectionism: how industry adjusts to competitive shocks,' *Journal of Business Strategy*, vol. 2(2), pp. 37–46.

Yeager, M. A. (1980), 'Trade protection as an international commodity: the case of steel,' *The Journal of Economic History*, vol. XL(1), pp. 33–42.

Journals and newspapers

American Metal Market
Business Week
Chilton's Iron Age
Consumer Reports
Department of Commerce News
Department of Treasury News
Edinburgh Review
Federal Register
Financial Times
Frankfurter Zeitung
Le Monde
Mercury News (San Jose, California)
New York Times
Official Journal of the European Communities
The Economist
The Statist
Wall Street Journal
Washington Post

INDEX

Numbers in italics refer to figures in the text